MATHEMATICS

Assessment

&

Intervention

in a PLC at Work®

Timothy D. Kanold   **Sarah Schuhl**

Matthew R. Larson   Bill Barnes

Jessica Kanold-McIntyre   Mona Toncheff

A Joint Publication With

Solution Tree | Press

a division of
Solution Tree

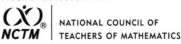

NATIONAL COUNCIL OF
TEACHERS OF MATHEMATICS

555 North Morton Street
Bloomington, IN 47404
800.733.6786 (toll free) / 812.336.7700
FAX: 812.336.7790

email: info@SolutionTree.com
SolutionTree.com

Visit **go.SolutionTree.com/MathematicsatWork** to download the free reproducibles in this book.

Printed in the United States of America

Library of Congress Cataloging-in-Publication Data

Names: Kanold, Timothy D., author.
Title: Mathematics assessment and intervention in a PLC at work / Timothy D.
   Kanold [and five others].
Other titles: Mathematics assessment and intervention in a professional
   learning community at work
Description: Bloomington, IN : Solution Tree Press, [2018] | Includes
   bibliographical references and index.
Identifiers: LCCN 2017046581 | ISBN 9781945349973 (perfect bound)
Subjects: LCSH: Mathematics--Study and teaching. | Mathematics
   teachers--Training of.
Classification: LCC QA11.2 .G7424 2018 | DDC 510.71/2--dc23 LC record available at https://lccn.
loc.gov/2017046581

Solution Tree
Jeffrey C. Jones, CEO
Edmund M. Ackerman, President

Solution Tree Press
President and Publisher: Douglas M. Rife
Editorial Director: Sarah Payne-Mills
Art Director: Rian Anderson
Managing Production Editor: Caroline Cascio
Senior Production Editor: Christine Hood
Senior Editor: Amy Rubenstein
Copy Editor: Ashante K. Thomas
Proofreader: Elisabeth Abrams
Text and Cover Designer: Laura Cox
Editorial Assistants: Jessi Finn and Kendra Slayton

Acknowledgments

Timothy D. Kanold

First and foremost, my thanks to each author on our team for understanding the joy, pain, and hard work of the writing journey, and for giving freely of their deep mathematics talent to so many others. Sarah, Matt, Bill, Mona, and especially my daughter, Jessica, you are each a gifted colleague and friend.

My thanks also to our reviewers, colleagues who have dedicated their lives to the work and effort described within the pages of this assessment book, and especially Kit, Sharon, and Jenn, who agreed to dedicate the time necessary to review each book in the series. I personally am grateful to each reviewer, and I know that you also believe that *every student can learn mathematics*.

Thanks, too, to Jeff Jones, Douglas Rife, Christine Hood, and the entire editorial team from Solution Tree for their belief in our vision and work in K–12 mathematics education and for making our writing and ideas so much better.

Thanks also to my wife, Susan, and the members of our "fambam" who understand how to love me and formatively guide me through the good and the tough times a series project as bold as this requires.

And finally, my thanks go to you, the reader. May you rediscover your love for mathematics and teaching each and every day. The story of your life work matters.

Sarah Schuhl

Working to advance mathematics education within the context of a professional learning community can be a daunting task. Many thanks to Tim for his vision, encouragement, feedback, and support. Since our paths first crossed, he has become my mentor, colleague, and friend. Your heart for students and educators is a beautiful part of all you do.

Additionally, thank you to my dear friends and colleagues Mona, Bill, Jessica, and Matt, led by Tim in his team approach to completing this series. I am grateful to work with such a talented team of educators who are focused on ensuring the mathematics learning of each and every student. You have challenged me and made the work both inspiring and fun.

To the many teachers I have worked with, thank you for your tireless dedication to students and your willingness to let me learn with you. A special thank you to the teachers at Aloha Huber Park School for working with me to give tasks so we could gather evidence of student reasoning and thinking to share.

Finally, I was only able to be a part of this series because of the support, love, and laughter given to me by my husband, Jon, and our sons, Jacob and Sam. Your patience and encouragement mean the world to me and will be forever appreciated.

Matthew R. Larson

It was an honor to be a member of an authorship team that truly understands each and every student can learn mathematics at a high level if the necessary conditions are in place. I thank all of you for your leadership and commitment to students and their teachers.

My thanks also go out to the thousands of dedicated teachers of mathematics who continue their own learning and work tirelessly every day to reach their students to ensure they experience mathematics in ways that prepare them for college and careers, promote active engagement in our democratic society, and help them experience the joy and wonder of learning mathematics. I hope this book will support you in your critical work.

Bill Barnes

First and foremost, thank you to our authorship team, Sarah, Mona, Jessica, and Matt, colleagues and friends who helped me overcome challenges associated with writing about the deep-rooted culture associated with homework and grading practices. Special thanks to Tim Kanold for continuing to believe in me, for supporting my work, and for helping me find my voice.

Special thanks to my wife, Page, and my daughter, Abby, who encourage me to pursue a diverse set of professional and personal interests. I am so grateful for your love, support, and patience as I continue to grow as a husband and father.

Finally, thanks to my work family, friends, and colleagues from the Howard County Public School System. I would like to especially thank our superintendent, Dr. Michael J. Martirano, for insisting that I follow my passion and continue to grow as a professional.

Jessica Kanold-McIntyre

I am extremely grateful to the writing team of Bill, Mona, Sarah, Tim, and Matt for their inspiration and collaboration through this writing process. Thank you to the teachers and leaders that have encouraged and influenced me along my own professional journey. Your passion and dedication to students and student learning continue to help me grow and learn. A special thanks to my husband for his support and encouragement throughout the process of writing this book. And, last but not least, I'd like to acknowledge our two dogs, Mac and Lulu, and my daughter Abigail Rose, who was born during the writing of this book, for their constant love.

Mona Toncheff

During my educational career, I have had the privilege of teaching students with diverse learning needs and backgrounds. Each year, my goal is to make mathematics more accessible to inspire students to continue their mathematics journey.

Now, I have the privilege of working with teachers and leaders across the nation whose goal is the same. The teachers and leaders with whom I have learned over the past twenty-five years have inspired this book series. The vision for a deeper understanding of mathematics for each and every student is achievable when we collectively respond to diverse student learning.

I am forever grateful to my husband, Gordon, and team Toncheff, who always support and encourage my professional pursuits.

Finally, thank you Tim for your leadership, mentorship, and vision for this series; and to Sarah, Matthew, Bill, and Jess; I am a better leader because of learning and leading with you.

Solution Tree Press would like to thank the following reviewers:

Kim Bailey
Educational Consultant
Mission Viejo, California

Kris Cunningham
Mathematics Content Specialist
Phoenix Union High School District
Phoenix, Arizona

Jennifer Deinhart
K–8 Mathematics Specialist
Mason Crest Elementary School
Annandale, Virginia

Kit Norris
Educational Consultant
Hudson, Massachusetts

Sharon Rendon
Mathematics Consultant
NCSM Central 2 Regional Director
Rapid City, South Dakota

Nick Resnick
Senior Program Manager
California Education Partners
San Francisco, California

John W. Staley
Director of Mathematics PreK–12
Baltimore County Public Schools
Towson, Maryland

Wendy Weatherwax
Secondary Mathematics Project Facilitator
Clark County School District
Las Vegas, Nevada

Gwen Zimmermann
Assistant Principal for Teaching and Learning
Adlai E. Stevenson High School
Lincolnshire, Illinois

Visit **go.SolutionTree.com/MathematicsatWork** to download the free reproducibles in this book.

Table of Contents

About the Authors

Timothy D. Kanold, PhD, is an award-winning educator, author, and consultant and national thought leader in mathematics. He is former director of mathematics and science and served as superintendent of Adlai E. Stevenson High School District 125, a model professional learning community (PLC) district in Lincolnshire, Illinois.

Dr. Kanold is committed to equity and excellence for students, faculty, and school administrators. He conducts highly motivational professional development leadership seminars worldwide with a focus on turning school vision into realized action that creates greater equity for students through the effective delivery of the PLC process by faculty and administrators.

He is a past president of the National Council of Supervisors of Mathematics (NCSM) and coauthor of several best-selling mathematics textbooks over several decades. Dr. Kanold has authored or coauthored thirteen books on K–12 mathematics and school leadership since 2011, including the bestselling book *HEART!* He also has served on writing commissions for the National Council of Teachers of Mathematics (NCTM) and has authored numerous articles and chapters on school leadership and development for education publications since 2006.

Dr. Kanold received the 2017 Ross Taylor/Glenn Gilbert Mathematics Education Leadership Award from the National Council of Supervisors of Mathematics,

the international 2010 Damen Award for outstanding contributions to the leadership field of education from Loyola University Chicago, 1986 Presidential Awards for Excellence in Mathematics and Science Teaching, and the 1994 Outstanding Administrator Award from the Illinois State Board of Education. He serves as an adjunct faculty member for the graduate school at Loyola University Chicago.

Dr. Kanold earned a bachelor's degree in education and a master's degree in mathematics from Illinois State University. He also completed a master's degree in educational administration at the University of Illinois and received a doctorate in educational leadership and counseling psychology from Loyola University Chicago.

To learn more about Timothy D. Kanold's work, visit his blog, *Turning Vision Into Action* (www.turning visionintoaction.today) and follow him @tkanold on Twitter.

Sarah Schuhl is an educational coach and consultant specializing in mathematics, professional learning communities, common formative and summative assessments, school improvement, and response to intervention (RTI). She has worked in schools as a secondary mathematics teacher, high school instructional coach, and K–12 mathematics specialist.

Schuhl was instrumental in the creation of a PLC in the Centennial School District in Oregon, helping teachers make large gains in student achievement. She earned the Centennial School District Triple C Award in 2012.

Sarah designs meaningful professional development in districts throughout the United States focused on strengthening the teaching and learning of mathematics, having teachers learn from one another when working effectively as a collaborative team in a PLC, and striving to ensure the learning of each and every student through assessment practices and intervention. Her practical approach includes working with teachers and administrators to implement assessments for learning, analyze data, collectively respond to student learning, and map standards.

Since 2015, Schuhl has co-authored the books *Engage in the Mathematical Practices: Strategies to Build Numeracy and Literacy With K–5 Learners* and *School Improvement for All: A How-to Guide for Doing the Right Work.*

Previously, Schuhl served as a member and chair of the National Council of Teachers of Mathematics (NCTM) editorial panel for the journal *Mathematics Teacher.* Her work with the Oregon Department of Education includes designing mathematics assessment items, test specifications and blueprints, and rubrics for achievement-level descriptors. She has also contributed as a writer to a middle school mathematics series and an elementary mathematics intervention program.

Schuhl earned a bachelor of science in mathematics from Eastern Oregon University and a master of science in mathematics education from Portland State University.

To learn more about Sarah Schuhl's work, follow her @SSchuhl on Twitter.

Matthew R. Larson, PhD, is an award-winning educator and author who served as the K–12 mathematics curriculum specialist for Lincoln Public Schools in Nebraska for more than twenty years. He served as president of the National Council of Teachers of Mathematics (NCTM) from 2016–2018. Dr. Larson has taught mathematics at the elementary through college levels and has held an honorary appointment as a visiting associate professor of mathematics education at Teachers College, Columbia University.

He is coauthor of several mathematics textbooks, professional books, and articles on mathematics education, and was a contributing writer on the influential publications *Principles to Actions: Ensuring Mathematical Success for All* (NCTM, 2014) and *Catalyzing Change in High School Mathematics: Initiating Critical Conversations* (NCTM, 2018). A frequent keynote speaker at national meetings, Dr. Larson's humorous presentations are well-known for their application of research findings to practice.

Dr. Larson earned a bachelor's degree and doctorate from the University of Nebraska–Lincoln, where he is an adjunct professor in the department of mathematics.

Bill Barnes is the chief academic officer for the Howard County Public School System in Maryland. He is also the second vice president of the NCSM and has served as an adjunct professor for Johns Hopkins University, the University of Maryland–Baltimore County, McDaniel College, and Towson University.

Barnes is passionate about ensuring equity and access in mathematics for students, families, and staff. His experiences drive his advocacy efforts as he works to ensure opportunity and access to underserved and underperforming populations. He fosters partnerships among schools, families, and community resources in an effort to eliminate traditional educational barriers.

A past president of the Maryland Council of Teachers of Mathematics, Barnes has served as the affiliate service committee eastern region 2 representative for the NCTM and regional team leader for the NCSM.

Barnes is the recipient of the 2003 Maryland Presidential Award for Excellence in Mathematics and Science Teaching. He was named Outstanding Middle School Math Teacher by the Maryland Council of Teachers of Mathematics and Maryland Public Television and Master Teacher of the Year by the National Teacher Training Institute.

Barnes earned a bachelor of science degree in mathematics from Towson University and a master of science degree in mathematics and science education from Johns Hopkins University.

To learn more about Bill Barnes's work, follow him @BillJBarnes on Twitter.

Jessica Kanold-McIntyre is an educational consultant and author committed to supporting teacher implementation of rigorous mathematics curriculum and assessment practices blended with research-informed instructional practices. She works with teachers and schools around the country to meet the needs of their students. Specifically, she specializes in building and supporting the collaborative teacher culture through the curriculum, assessment, and instruction cycle.

She has served as a middle school principal, assistant principal, and mathematics teacher and leader. As principal of Aptakisic Junior High School in Buffalo Grove, Illinois, she supported her teachers in implementing initiatives, such as the Illinois Learning Standards; Next Generation Science Standards; and the Social Studies College, Career, and Civic Life Framework for Social Studies State Standards, while also supporting a one-to-one iPad environment for all students. She focused on teacher instruction through the PLC process, creating learning opportunities around formative assessment practices, data analysis, and student engagement. She previously served as assistant principal at Aptakisic, where she led and supported special education, response to intervention (RTI), and English learner staff through the PLC process.

As a mathematics teacher and leader, Kanold-McIntyre strived to create equitable and rigorous learning opportunities for all students while also providing them with cutting-edge 21st century experiences that engage and challenge them. As a mathematics leader, she developed and implemented a districtwide process for the Common Core State Standards in Illinois and led a collaborative process to create mathematics curriculum guides for K–8 mathematics, algebra 1, and algebra 2. She currently serves as a board member for the National Council of Supervisors of Mathematics (NCSM).

Kanold-McIntyre earned a bachelor's degree in elementary education from Wheaton College and a master's degree in educational administration from Northern Illinois University.

To learn more about Jessica Kanold-McIntyre's work, follow her @jkanold on Twitter.

Mona Toncheff, an educational consultant and author, is also currently working as a project manager for the Arizona Mathematics Partnership (a National Science Foundation–funded grant). A passionate educator working with diverse populations in a Title I district, she previously worked as both a mathematics teacher and a mathematics content specialist for the Phoenix Union High School District in Arizona. In the latter role, she provided professional development to high school teachers and administrators related to quality mathematics teaching and learning and working in effective collaborative teams.

As a writer and consultant, Mona works with educators and leaders nationwide to build collaborative teams, empowering them with effective strategies for aligning curriculum, instruction, and assessment to ensure all students receive high-quality mathematics instruction.

Toncheff is currently an active member of the National Council of Supervisors of Mathematics (NCSM) board and has served NCSM in the roles of secretary (2007–2008), director of western region 1 (2012–2015), second vice president (2015–2016), first vice president (2016–2017), marketing and e-news editor (2017–2018), and president-elect (2018–2019). In addition to her work with NCSM, Mona is also the current president of Arizona Mathematics Leaders (2016–2018). She was named 2009 Phoenix Union High School District Teacher of the Year; and in 2014, she received the Copper Apple Award for leadership in mathematics from the Arizona Association of Teachers of Mathematics.

Toncheff earned a bachelor of science degree from Arizona State University and a master of education degree in educational leadership from Northern Arizona University.

To learn more about Mona Toncheff's work, follow her @toncheff5 on Twitter.

To book Timothy D. Kanold, Sarah Schuhl, Matthew R. Larson, Bill Barnes, Jessica Kanold-McIntyre, or Mona Toncheff for professional development, contact pd@SolutionTree.com.

Preface

By Timothy D. Kanold

In the early 1990s, I had the honor of working with Rick DuFour at Adlai E. Stevenson High School in Lincolnshire, Illinois. During that time, Rick—then principal of Stevenson—began his revolutionary work as one of the architects of the Professional Learning Communities at Work® (PLC) process. My role at Stevenson was to initiate and incorporate the elements of the PLC process into the K–12 mathematics programs, including the K–5 and 6–8 schools feeding into the Stevenson district.

In those early days of our PLC work, we understood the grade-level or course-based mathematics collaborative teacher teams provided us a chance to share and become more transparent with one another. We exchanged knowledge about our own growth and improvement as teachers and began to create and enhance student agency for learning mathematics. As colleagues and team members, we taught, coached, and learned from one another.

And yet, we did not know our mathematics *assessment* story. We did not have much clarity on an assessment vision that would improve student learning in mathematics.

We did not initially understand how the work of our collaborative teacher teams—especially in mathematics at all grade levels—when focused on the right assessment actions, could erase inequities in student learning that the wide variance of our professional assessment practice caused.

Through our work together, we realized that, without intending to, we often were creating massive gaps in student learning because of our isolation from one another; our isolated decisions about the specifics of teaching mathematics were a crushing consequence in a vertically connected curriculum like mathematics.

We also could not have anticipated one of the best benefits of working in community with one another: the benefit of belonging to something larger than ourselves. There is a benefit to learning about various teaching and assessing strategies from each other, *as professionals*. We realized it was often in community that we find deeper meaning to our work, and strength in the journey as we solve the complex issues we faced each day and each week of the school season, *together*.

As we began our collaborative mathematics work at Stevenson and with our feeder districts, we discovered quite a bit about our mathematics teaching. And remember, we were doing this work together in the early 1990s—well before the ideas of transparency in practice and observing and learning from one another in our professional work became as popular as they are today.

We discovered that if we were to become a professional *learning* community, then experimenting together with effective practices needed to become our norm. We needed to learn more about the mathematics curriculum, the nature of the mathematics tasks we were choosing each day, and the types of lessons and assessments we were developing and using. And, we realized we needed to do so *together*.

The idea of collaborative focus to the real work we do as mathematics teachers is at the heart of the *Every Student Can Learn Mathematics* series. The belief, that if we do the right work together, then just maybe every student can learn mathematics, has been the driving force of our work for more than thirty years. And thus, the title of this mathematics professional development series was born.

In this series, we emphasize the concept of *team action*. We recognize some readers may be the only members of a grade level or mathematics course. In that case, we recommend you work with a colleague at a grade level or course above or below your own. Or, work with other job-alike teachers across a geographical region as technology allows. Collaborative teams are the engines that drive the PLC process.

A PLC in its truest form is "an ongoing process in which educators work collaboratively in recurring cycles of collective inquiry and action research to achieve better results for the students they serve" (DuFour, DuFour, Eaker, Many, & Mattos, 2016, p. 10). This book and the other three in the *Every Student Can Learn Mathematics* series feature a wide range of voices, tools, and discussion protocols offering advice, tips, and knowledge for your PLC-based collaborative mathematics team.

The coauthors of the *Every Student Can Learn Mathematics* series—Bill Barnes, Jessica Kanold-McIntyre, Matt Larson, Sarah Schuhl, and Mona Toncheff—have each been on their own journeys with the deep and collaborative work of PLCs for mathematics. They have all spent significant time in the classroom as highly successful practitioners, leaders, and coaches of K–12 mathematics teams, designing and leading the structures and the culture necessary for effective and collaborative team efforts. They have lived through and led the mathematics professional growth actions this book advocates within diverse K–12 educational settings in rural, urban, and suburban schools.

In this book, we tell our mathematics assessment story. It is a story of developing high-quality assessments for your students, scoring those assessments with fidelity, and helping students use those assessments for formative learning. It is a K–12 story that when well implemented, will bring great satisfaction to your work as a mathematics professional and result in a positive impact on your students.

We hope you will join us in the journey of significantly improving student learning in mathematics by leading and improving your assessment story for your team, your school, and your district. The conditions and the actions for adult learning of mathematics *together* are included in these pages. We hope the stories we tell, the tools we provide, and the opportunities for reflection serve you well in your daily work in a discipline we all love—mathematics!

Introduction

If you are a teacher of mathematics, then this book is for you! Whether you are a novice or a master teacher; an elementary, middle, or high school teacher; a rural, suburban, or urban teacher; this book is for you. It is for all teachers and support professionals who are part of the K–12 mathematics learning experience.

Teaching mathematics so *each and every student* learns the K–12 college-preparatory mathematics curriculum, develops a positive mathematics identity, and becomes empowered by mathematics is a complex and challenging task. Trying to solve that task in isolation from your colleagues will not result in erasing inequities that exist in your schools. The pursuit and hope of developing into a collaborative community with your colleagues and moving away from isolated professional practice are necessary, hard, exhausting, and sometimes overwhelming.

Your professional life as a mathematics teacher is not easy. In this book, you and your colleagues will focus your time and energy on collaborative efforts that result in significant improvement in student learning.

Some educators may ask, "Why become engaged in collaborative mathematics teaching actions in your school or department?" The answer is simple: *equity*.

What is equity? To answer that, it is helpful to first examine inequity. In traditional schools in which teachers work in isolation, there is often a wide discrepancy in teacher practice. Teachers in the same grade level or course may teach and assess mathematics quite differently—there may be a lack of consistency in what teachers expect students to know and be able to do,

how they will know when students have learned, what they will do when students have not learned, and how they will proceed when students have demonstrated learning. Such wide variance in potential teacher practice among grade-level and course-based teachers then causes inequities as students pass from course to course and grade to grade.

These types of equity issues require you and your colleagues to engage in team discussions around the development and use of assessments that provide evidence of and strategies for improving student learning.

Equity and PLCs

The PLC at Work process is one of the best and most promising models your school or district can use to build a more equitable response for student learning. The architects of the PLC process, Richard DuFour, Robert Eaker, and Rebecca DuFour, designed the process around three big ideas and four critical questions that placed learning, collaboration, and results at the forefront of our work (DuFour, et al., 2016). Schools and districts that commit to the PLC transformation process rally around the following three big ideas (DuFour et al., 2016).

1. **A focus on learning:** Teachers focus on learning as the fundamental purpose of the school rather than on teaching as the fundamental purpose.

2. **A collaborative culture:** Teachers work together in teams interdependently to achieve a

common goal or goals for which members are mutually accountable.

3. **A results orientation:** Team members are constantly seeking evidence of the results they desire—high levels of student learning.

Additionally, teacher teams within a PLC focus on four critical questions (DuFour et al., 2016).

1. What do we want all students to know and be able to do?

2. How will we know if students learn it?

3. How will we respond when some students do not learn?

4. How will we extend the learning for students who are already proficient?

The four critical PLC questions provide an equity lens for your professional work. Imagine the opportunity gaps that will exist if you and your colleagues do not agree on the level of rigor for PLC critical question 1 (DuFour et al., 2016): What do we want all students to know and be able to do?

Imagine the devastating effects on K–12 students if you do not reach complete team agreement on the high-quality criteria for the assessments you administer (see PLC critical question 2) and your routines for how you score them. Imagine the lack of student agency (their voice in learning) if you do not work together to create a unified, robust formative mathematics assessment process for helping students *own* their response when they are and are not learning.

To answer these four PLC critical questions well requires structure through the development of *products* for your work together, and a formative culture through the *process* of how you work with your team to *use* those products.

The concept of your *team reflecting together and then taking action* around the right work is an emphasis in the *Every Student Can Learn Mathematics* series.

The Reflect, Refine, and Act Cycle

Figure I.1 illustrates the reflect, refine, and act cycle, our perspective about the process of lifelong learning—for us, for you, and for your students. The very nature of our profession—education—is about the development of skills toward learning. Those skills are part of an ongoing process we pursue together.

More important, the reflect, refine, and act cycle is a *formative* learning cycle we describe throughout all four books in the series. When you embrace mathematics learning as a *process*, you and your students:

- **Reflect:** Work the task, and then ask, "Is this the best solution strategy?"

- **Refine:** Receive FAST feedback and ask, "Do I embrace my errors?"

- **Act:** Persevere and ask, "Do I seek to understand my own learning?"

The intent of this *Every Child Can Learn Mathematics* series is to provide you with a systemic way to structure and facilitate deep team discussions necessary to lead an effective and ongoing adult and student learning process each and every school year.

Mathematics in a PLC at Work Framework

The *Every Student Can Learn Mathematics* series includes four books that focus on a total of six teacher team actions and two coaching actions within four larger categories.

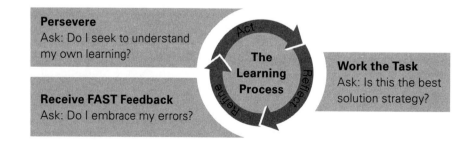

Figure I.1: Reflect, refine, and act cycle for formative student learning.

1. *Mathematics Assessment and Intervention in a PLC at Work*

2. *Mathematics Instruction and Tasks in a PLC at Work*

3. *Mathematics Homework and Grading in a PLC at Work*

4. *Mathematics Coaching and Collaboration in a PLC at Work*

Figure I.2 shows each of these four categories and the two actions within it. These eight actions focus on the nature of the ongoing, unit-by-unit professional work of your teacher teams and how you should respond to the four critical questions of a PLC (DuFour et al., 2016).

Most commonly, a collaborative team consists of two or more teachers who teach the same grade level or course. Through your focused work addressing the four

Every Student Can Learn Mathematics series' Team and Coaching Actions Serving the Four Critical Questions of a PLC at Work	1. What do we want all students to know and be able to do?	2. How will we know if students learn it?	3. How will we respond when some students do not learn?	4. How will we extend the learning for students who are already proficient?
Mathematics Assessment and Intervention in a PLC at Work				
Team action 1: Develop high-quality common assessments for the agreed-on essential learning standards.	■	■		
Team action 2: Use common assessments for formative student learning and intervention.			■	■
Mathematics Instruction and Tasks in a PLC at Work				
Team action 3: Develop high-quality mathematics lessons for daily instruction.	■	■		
Team action 4: Use effective lesson designs to provide formative feedback and student perseverance.			■	■
Mathematics Homework and Grading in a PLC at Work				
Team action 5: Develop and use high-quality common independent practice assignments for formative student learning.	■	■		
Team action 6: Develop and use high-quality common grading components and formative grading routines.			■	■
Mathematics Coaching and Collaboration in a PLC at Work				
Coaching action 1: Develop PLC structures for effective teacher team engagement, transparency, and action.	■	■		
Coaching action 2: Use common assessments and lesson-design elements for teacher team reflection, data analysis, and subsequent action.			■	■

Figure I.2: Mathematics in a PLC at Work framework.

Visit go.SolutionTree.com/MathematicsatWork for a free reproducible version of this figure.

critical questions of a PLC (DuFour et al., 2016), you provide every student in your grade level or course with equitable learning experiences and expectations, opportunities for sustained perseverance, and robust formative feedback, regardless of the teacher he or she receives.

If, however, you are a singleton (a lone teacher who does not have a colleague who teaches the same grade level or course), you will have to determine who it makes the most sense for you to work with as you strengthen your lesson design and student feedback skills. Leadership consultant and author Aaron Hansen (2015) suggests the following possibilities for creating teams for singletons.

- Vertical teams (for example, a primary school team of grades K–2 teachers or a middle school mathematics department team for grades 6–8)

- Virtual teams (for example, a team comprising teachers from different sites who teach the same grade level or course and collaborate virtually with one another across geographical regions)

- Grade-level or course-based teams (for example, a team of grade-level or course-based teachers in which each teacher teaches all sections of grade 6, grade 7, or grade 8; the teachers might expand to teach and share two or three grade levels instead of only one to create grade-level or course-based teams)

About This Book

Every grade-level or course-based collaborative team in a PLC culture is expected to meet on an ongoing basis to discuss how its mathematics lessons and assessments ask and answer the four critical questions as students are learning (DuFour et al., 2016). For this book in the series, we explore two specific *assessment team actions* for your professional work.

- **Team action 1:** Develop high-quality common assessments for the agreed-on essential learning standards.

- **Team action 2:** Use common assessments for formative student learning and intervention.

This book emphasizes intentional differentiation between the mathematics assessment instruments or *products* your team develops and produces (part 1, chapters 1–3) and the *process* (part 2, chapters 4–6) for how your team uses those products to analyze the student learning results of your mathematics work week after week. Part 1 examines the quality of common mathematics assessment instruments you use during and at the end of each unit of study. You will evaluate your current common assessments (quizzes and tests) based on eight mathematics assessment design criteria, including the calibration of your scoring for those assessments. Part 2 explores how to use the high-quality common assessment instruments you develop to enhance and improve the student learning process.

The tools and protocols in this book are designed to help you become confident and comfortable in mathematics assessment conversations with one another, and move toward greater transparency in your assessment practices with colleagues.

In this book, you will find discussion tools that offer questions for reflection, discussion, and action about your mathematics assessment practices and routines. We invite you to write your personal assessment story in the *teacher reflection* boxes as well. You will also find some personal stories from the authors of this series. These stories provide a glimpse into the authors' personal insights, experiences, and practical advice connected to some of the strategies and ideas in this book.

The *Every Student Can Learn Mathematics* series is steeped in the belief that as classroom teachers of mathematics, your decisions and your daily actions matter. You have the power to decide and choose the mathematical tasks students are required to complete during lessons, homework, unit assessments such as quizzes and tests, and the projects and other high-performance tasks you design. You have the power to determine the rigor for those mathematical tasks, the nature of student communication and discourse to learn those tasks, and whether or not learning mathematics should be a formative process for you and your students.

Visit **go.SolutionTree.com/MathematicsatWork** for the free reproducibles "Unit Samples for Essential Learning Standards in Mathematics," which provides example learning targets aligned to unit standards for assessment and reflection, and "Online Resources Reference Guide for Mathematics Support," which offers a comprehensive list of free online resources to support your work in mathematics teaching and learning.

As you embrace the belief that together, the work of your professional learning community can overcome the complexities of your work and obstacles to learning you face each day, then *every student can learn mathematics* just may become a reality in your school.

Team Action 1:
Develop High-Quality Common Assessments for the Agreed-On Essential Learning Standards

*To begin with the end in mind means to start with a clear understanding of your destination.
It means to know where you're going so that you better understand
where you are now so that the steps you take are always in the right direction.*

—*Stephen R. Covey*

During your career, you can use common mathematics assessments to create a more equitable learning experience for every student, inspire student learning, and help students persevere in their mathematics learning based on the feedback they receive and their response to that feedback. Part 1 of this book examines the criteria for evaluating and developing high-quality, unit-by-unit assessments (both during and at the end of a unit) and the accuracy of the scoring protocols used for those assessments.

There are several team discussion tools designed to support your team's assessment work in part 1.

- Common Assessment Self-Reflection Protocol (figure 1.1, page 11)
- Assessment Instrument Quality Evaluation Rubric (figure 2.1, page 14)
- High-Quality Assessment Evaluation (figure 2.2, page 15)
- Sample Assessment Questions for Team Scoring (figure 2.5, page 29)
- Scoring Assessment Tasks (figure 2.6, page 30)
- Sample Grade 4 End-of-Unit Assessment (figure 3.3, page 43)
- Sample High School Algebra 1/Integrated Mathematics I End-of-Unit Assessment (figure 3.4, page 47)
- Assessment Instrument Alignment and Scoring Rubric (figure 3.8, page 61)

Mathematics assessment throughout the year becomes an opportunity for students to *reflect*, *refine*, and *act* both during and at the end of a unit. Your students should be expected to use the assessments and feedback you provide to reflect on their understanding for each mathematics standard, and then act to refine their errors in the process. This formative feedback process begins with an examination of the quality of your common unit mathematics assessments with your colleagues.

CHAPTER 1

The Purpose and Benefit of Common Mathematics Assessments

When you're surrounded by people who share a passionate commitment around a common purpose, anything is possible.

—*Howard Schultz*

Chances are, you already use during-the-unit and end-of-unit assessments to measure and record student learning. You may or may not know the routines your colleagues use to ensure the assessments you create are in common. You also may not have a common strategy for assessing the quality of those assessments.

In *Learning by Doing*, authors DuFour, DuFour, Eaker, Many, and Mattos (2016) indicate that common assessments:

- Promote efficiency for teachers
- Promote equity for students
- Provide an effective strategy for determining whether the guaranteed curriculum is being taught and, more importantly, learned
- Inform the practice of individual teachers
- Build a team's capacity to improve its program
- Facilitate a systematic, collective response to students who are experiencing difficulty
- Offer the most powerful tool for changing the adult behavior and practice (p. 149)

Reflect for a moment about the student's point of view regarding the mathematics assessments they receive throughout the school year. In a PLC culture, the student mathematics assessment experience should be the same, regardless of the teacher the school assigns. The rigor of the mathematical tasks for each essential learning standard, the format of the test questions, the number of test questions, the alignment of those questions to each standard, the strategies students use to demonstrate learning, and the scoring of those assessment questions all depend on your willingness to work together and erase any potential inequities that a lack of transparency with colleagues can cause.

Working together also means creating common mathematics assessments so students experience the same mathematical rigor expectations across classrooms. In a vertically connected curriculum like mathematics, the equitable use of common unit assessments is more likely to affect the student K–12 learning experience. The purposes behind common assessments are significant.

Four Specific Purposes

There are four specific purposes supporting the use of common mathematics assessments.

1. Improved equity and opportunity for student learning

2. Professional development

3. Student ownership of learning

4. Student intervention and support from the teacher team

Improved Equity and Opportunity for Student Learning

Essentially, you minimize the wide variance in student expectations between you and your colleagues

---

Personal Story 66 **SARAH SCHUHL**

When I first began working as part of a professional learning community at Centennial High School, we began our assessment work as a team in geometry. We discovered some teachers heavily weighted proofs and required students to solve quadratic equations when performing mathematical tasks related to standards for special angles.

Other teachers on our team only require fill-in-the-blank proofs using less complex linear equations for the same standard. This meant we were using variant levels of rigor for the same standard and equitable learning experiences, and expectations for our students did *not* exist in geometry.

Similarly, when working as an instructional coach and looking at an assessment related to rounding with my fourth-grade colleagues, we noticed that some of the teachers only asked students to round to the nearest ten, others required explanations, and still others asked questions for conceptual understanding, such as, What is the largest number that could be rounded to fifty when rounding to the nearest ten?

(an inequity creator) when you work collaboratively to design high-quality assessment instruments appropriate to the identified essential learning standards for the unit. Consider the two equity experiences discussed by author Sarah Schuhl.

Common mathematics unit assessments then, are a first order of business for your professional work life. They are at the heart of working effectively and more equitably as your collaborative team answers the second critical question of the PLC culture (DuFour et al., 2016): How will we know if they learn it? Agreement on the mathematical tasks used to assess each essential learning standard for a unit of study is where team collaboration often begins.

Professional Development

When working collaboratively to design high-quality mathematics assessments, you learn from colleagues and gain clarity as to the rigor of the tasks for each essential standard of a mathematics unit. This expectation often requires team discussion on the level of cognitive demand for each mathematics task or problem on the assessment and the appropriate amount of application or mathematical modeling. As your team asks, "Do the tasks (the assessment questions) align to the expectations of the learning standards?" you begin a professional development process about the mathematics content.

You build the capacity of your team to learn from one another as you share insights into the common assessment tasks and expectations for students to receive full credit in response to any mathematics problem on the assessment.

TEACHER *Reflection* ✏

How do you currently determine if the quality of the mathematical tasks (the test questions) you use to assess student learning is appropriate to the understanding levels expected by each essential learning standard of the unit?

_____

_____

_____

_____

_____

_____

_____

_____

_____

_____

_____

Student Ownership of Learning

As a result of common mathematics assessments, you and your colleagues can help students identify which mathematics standards they have learned or not learned *yet* and work together to find appropriate and common responses to interventions or extensions (discussed in part 2 of this book). Your students begin to take greater ownership in their learning of the essential mathematics standards as they respond to evidence of learning from the common unit assessments.

As your students learn to use daily student trackers to monitor their progress on each essential learning standard, they can begin to understand or view mathematics assessment not as something to be feared, but rather as an integral aspect of their learning process. You can find sample student trackers in chapter 6 on pages 98–103.

Student Intervention and Support From the Teacher Team

Student support systems for mathematics should rely on the essential learning standards for the unit. When during- and end-of-unit assessments are common, it allows each member of your team to help any students in the grade level or course you teach. If a student is not sure about a specific learning standard and needs help, he or she can receive help from any teacher on your team, and each teacher can anticipate the usual trouble spots that may occur. The intentional and targeted mathematics intervention (discussed in chapter 5) becomes a more shared, unified, engaging, and equitable learning experience for each student.

TEAM RECOMMENDATION

Know the Purpose of Common Assessments

- Understand how common assessments erase inequities in student learning and allow for teacher learning, opportunities for students to identify what they have learned and not learned, and a collective response to re-engaging students in learning.

- Make a team commitment to the development of common end-of-unit and mid-unit assessments throughout the year.

The purposes of common assessments reveal that your team's actions toward creating a common mathematics assessment tool are a necessary part of your professional work, but not sufficient to fully impact student learning. It is only the beginning of a full formative feedback and assessment process for you, your colleagues, and your students.

The Assessment Instrument Versus the Assessment Process

Not only should you and your colleagues create the common assessment instrument itself (the actual quiz or test for each unit), you also should establish a process in which you and your colleagues use this assessment with students. This distinguishes between formative assessment *processes* you use and the assessment *instruments* you design as part of the formative learning process. DuFour et al. (2016) describe the importance of having common instruments and process this way:

> One of the most powerful, high-leverage strategies for improving student learning available to schools is the creation of frequent, high-quality common assessments by teachers who are working collaboratively to help a group of students acquire agreed-on knowledge and skills. (p. 141)

W. James Popham (2011) provides an analogy to describe the difference between summative assessment instruments (such as your end-of-unit tests) and formative assessment processes (such as what your students *do* in each lesson to learn the standards or *do* as a response to those test results). He describes the difference between a surfboard and surfing (Popham, 2011):

> While a surfboard represents an important tool in surfing, it is only that—a part of the surfing process. The entire process involves the surfer paddling out to an appropriate offshore location, selecting the right wave, choosing the most propitious moment to catch the chosen wave, standing upright on the surfboard, and staying upright while a curling wave rumbles toward shore. (p. 36)

The surfboard (the test) is a key component of the surfing process, but it is not the entire process.

Sarah's story illustrates NCTM's (2014) *Principles to Actions*, which adds, "Thinking of assessment as limited to 'testing' student learning rather than as a process that can advance it has been an obstacle to the effective use of assessment processes for decades" (p. 91). The mathematics assessment instruments your team creates are the tools your team uses to collect data about student demonstrations of the essential learning standards. The assessment instruments subsequently inform you and your students' ongoing decisions about learning mathematics throughout the school year.

Assessment instruments vary and can include tools such as class assignments, exit slips, journal entries, quizzes, unit tests, projects, and performance tasks, to name a few. However, to avoid inequities in the level of rigor for student work, and to serve the formative learning process, these assessment instruments must be *in common* for every teacher on your grade-level or course-based team.

Richard Stiggins (2007) notes, "You can enhance or destroy students' desire to succeed in school more quickly and permanently through your use of assessment than with any other tools you have at your disposal" (p. 22). *Enhance* and *destroy* are two powerfully charged words. You can probably think of a time when taking an assessment was a good experience and an awful experience—there are emotions tied to different types of assessment practices. How, then, can your team design and use assessments to enhance and promote

student learning? This book seeks to support your team in answering this essential question.

Richard DuFour (2015) further states:

> When members of a team work together to consider how they will approach assessing the skills and knowledge of their students, when every item or performance task is vetted by the entire team before it becomes part of the assessment, there is a much greater likelihood that the end result is a quality assessment compared to assessments that isolated teachers create. (p. 175)

Team-developed common mathematics assessments can create the continuity and equity needed to improve student learning. The first step in successfully collaborating to create these common assessments is to reflect upon and share, as a team, your current individual mathematics assessment practices.

Reflection on Team Assessment Practices

In order to create high-quality common mathematics assessment instruments, you and your colleagues first need to understand the purposes of the assessments you administer to students. The questions in figure 1.1 will help you and your team understand one another's perspectives related to common assessments, particularly about the design, value, and uses of common

---

Personal Story **SARAH SCHUHL**

Not all teaching experiences are the same. My first mathematics teaching job was at a small rural school serving grades 7–12 in eastern Oregon. Not only was I the mathematics teacher, I *was* the mathematics department. As such, I dutifully created and scored mid-unit quizzes and chapter tests for each course—just as I had experienced as a student many years before. Occasionally, I wondered if the assessments were good only to quickly convince myself that since I had taken questions from the publisher resources, they must be fine.

Fast forward to my second mathematics teaching position outside of Portland, Oregon, where teachers worked in teams and gave the same common assessment—mid-unit checks and end-of-unit tests. This was an eye-opening experience for me. I quickly learned the efficiency of collaboratively planning an assessment to make clear the learning expectations in each unit and ensure the results matched the learning expectations for student instruction. Incredible! Who knew that creating a common unit assessment could impact and grow student learning, not just measure and record it? This in turn grew my own understanding of the content, which then strengthened my daily instruction.

assessments related to student learning, and you'll begin to frame the work ahead in this book.

Use the self-reflection protocol in figure 1.1 to ask each member of your mathematics team about his or her personal unit-by-unit assessment practices. You can use team members' responses to find some initial common ground as you share your current mathematics assessment work.

Discuss your responses as a collaborative team to determine how similar or different your practices are as you build your team consensus for common assessment development.

Used well, common assessment instruments (your surfboards, so to speak) provide quality direction to you, your colleagues, and your students. The design of your unit assessments should also support your team's important assessment work. Consider your team response to the first two critical questions of a PLC (DuFour et al., 2016).

1. What do we want all students to know and be able to do?

2. How will we know if students learn it?

Directions: Use the following prompts to guide a team discussion of your common assessment practices for each unit. Share your results with other teams as needed.

Purpose of common assessments:

1. Why do we need common mathematics assessments for each unit?

Design of common mathematics assessments:

2. Do we write the essential learning standards on our tests? Who currently identifies and decides the essential learning standards for each unit?

3. Which assessments are common for our team (for example, quizzes, exit slips, mid-unit assessments, projects, and end-of-unit assessments)?

4. Who creates the assessments and the mathematics tasks (questions) for each assessment?

5. How do we organize the assessments (for example, by assessment type, or by essential learning standard)?

6. Do we expect students to show their work on the assessment? What work do we expect?

7. Are there common scoring agreements for each mathematics task (problem) on the assessment?

Use of common mathematics assessments:

8. Are students currently required to respond to errors made after taking common assessments?

9. What percent of a student's final grade is based on the result of a common assessment?

10. How does our team currently analyze data from common assessments to inform instructional decisions in the next unit?

Figure 1.1: Team discussion tool—Common assessment self-reflection protocol.

*Visit **go.SolutionTree.com/MathematicsatWork** for a free reproducible version of this figure.*

TEACHER *Reflection*

Reflect with your colleagues on the current process you use to design unit assessments.

1. Who writes the assessments: each teacher on the team, the central office, copied and administered from publisher resources, or some other entity?

2. What sources do you use?

3. How do you make decisions about scoring each mathematical task or question on the assessment?

_____

_____

_____

_____

_____

_____

_____

_____

_____

A proper team response to these two questions will be revealed through the mathematics assessments you and your team design for each unit throughout the school year. The first order of business for any collaborative team is to evaluate the quality of its current school- or district-driven mathematics assessments.

CHAPTER 2

Quality Common Mathematics Unit Assessments

Design is a funny word. Some people think design means how it looks.
But of course, if you dig deeper, it's really how it works.

—*Steve Jobs*

How do you decide if the unit-by-unit common mathematics assessment instruments you design are high quality? This is a question every teacher and leader of mathematics should ask. Tim Kanold describes his experience at Stevenson HSD 125 when he first asked this question of the middle school mathematics teachers from one of the Stevenson feeder districts.

The mathematics assessment work Tim describes created an early beta model for a test evaluation tool he and his fellow teachers could use to evaluate the quality of their unit-by-unit mathematics assessments. Figure 2.1 (page 14) provides a much deeper

and robust during-the-unit or end-of-unit assessment instrument evaluation tool your collaborative team can use to evaluate quality and build new and revised unit assessment instruments.

Your collaborative team should rate and evaluate the quality of its most recent end-of-unit or chapter assessment instruments (tests and quizzes) using the tools in figure 2.1 and figure 2.2 (page 15). How does your team's end-of-unit assessment score? Do you rate your current tests a twelve, sixteen, or twenty-two? How close does your assessment instrument come to scoring a twenty-seven or higher out of the thirty-two points possible in the test-quality protocol?

Personal Story TIMOTHY KANOLD

During my fourth year as the director of mathematics at Stevenson, it was clear to me that we were not very assessment literate. By this I mean we had very little knowledge about each other's assessment routines and practices. I remember collecting all the mathematics tests being given to our grades 6–8 feeder district middle school students during the months of October and November, and feeling a certain amount of dismay at the wide range of rigor, the lack of direction for how the assessments were organized, and our lack of clarity about the nature of the mathematics tasks being chosen for the exam.

In some instances, the questions we were asking sixth-grade students were more rigorous than the questions we were asking eighth-grade students. We had little or no shared direction for this most important aspect of our professional life. We needed to do a deep dive into the research and the expectations of high-quality mathematics assessments. It began a journey for all of our mathematics teachers to not only become assessment literate, but to use our mathematics assessments as a way to sustain student effort and learning at the right level of rigor across the grades.

High-Quality Assessment Criteria	Description of Level 1	Requirements of the Indicator Are Not Present	Limited Requirements of the Indicator Are Present	Substantially Meets the Requirements of the Indicator	Fully Achieves the Requirements of the Indicator	Description of Level 4
1. Identification and emphasis on essential learning standards (student-friendly language)	Essential learning standards are unclear, absent from the assessment instrument, or both. Some of the mathematical tasks (questions) may not align to the essential learning standards of the unit. The organization of assessment tasks is not clear.	1	2	3	4	Essential learning standards are clear, included on the assessment, and connected to the assessment tasks (questions).
2. Balance of higher- and lower-level-cognitive-demand mathematical tasks	Emphasis is on procedural knowledge with minimal higher-level-cognitive-demand mathematical tasks for demonstration of understanding.	1	2	3	4	Test is rigor balanced with higher-level and lower-level-cognitive-demand mathematical tasks present and aligned to the essential learning standards.
3. Variety of assessment-task formats and use of technology	Assessment contains only one type of questioning strategy—selected response or constructed response. There is little to no modeling of mathematics or use of tools. Use of technology (such as calculators) is not clear.	1	2	3	4	Assessment includes a blend of assessment types and modeling tasks or use of tools. Use of technology (such as calculators) is clear.
4. Appropriate and clear scoring rubric (points assigned or proficiency scale)	Scoring rubric is not evident or is inappropriate for the assessment tasks.	1	2	3	4	Scoring rubric is clearly stated and appropriate for each mathematical task.
5. Clarity of directions	Directions are missing or unclear. Directions are confusing for students.	1	2	3	4	Directions are appropriate and clear.
6. Academic language	Wording is vague or misleading. Academic language (vocabulary and notation) are not precise, causing a struggle for student understanding and access.	1	2	3	4	Academic language (vocabulary and notation) in tasks is direct, fair, accessible, and clearly understood by students. Teachers expect students to attend to precision in response.
7. Visual presentation	Assessment instrument is sloppy, disorganized, difficult to read, and offers no room for student work.	1	2	3	4	Assessment is neat, organized, easy to read, and well-spaced, with room for student work. There is also room for teacher feedback.
8. Time allotment	Few students can complete the assessment in the time allowed.	1	2	3	4	Students can successfully complete the assessment in the time allowed.

Figure 2.1: Team discussion tool—Assessment instrument quality evaluation rubric.

Visit go.SolutionTree.com/MathematicsatWork for a free reproducible version of this figure.

Directions: Examine your most recent common end-of-unit assessment, and evaluate its quality against the following eight criteria in figure 2.1 (page 14). Write your responses to each question in the following spaces.

1. Are the essential learning standards written on the assessment?

Discuss: What do our students think about learning mathematics? Do they think learning mathematics is about doing a bunch of random mathematics problems? Or, can they explain the essential learning standards in student-friendly *I can...* statements for each group of questions? Can they solve any mathematical tasks that might reflect a demonstration of learning that standard? Are our students able to use the essential learning standards and tasks to determine what they have learned and what they have not learned yet?

Note: In order for students to respond to the end-of-unit assessment feedback when the teacher passes it back, this is a necessary assessment feature.

2. Is there an appropriate balance of higher- and lower-level-cognitive-demand mathematical tasks on the assessment?

Discuss: What percentage of all tasks or problems on the assessment instrument is of lower-level cognitive demand? What percentage is of higher-level cognitive demand? Is there an appropriate balance? Is balancing rigor a major focus of our work?

Note: Use figure 2.4 (page 23) as a tool to determine the rigor. This will help you to better understand the level of cognitive demand. Also, see the appendix (page 111) for more advice on this criterion. As a good rule of thumb, the rigor-balance ratio should be about 30/70 (lower- to higher-level cognitive demand) on the assessment as appropriate to the standards on the assessment.

3. Is there variety of assessment formats and a clear use of technology?

Discuss: Does our assessment use a blend of assessment formats or types? If we use multiple choice, do we include questions with multiple possible answers? Do we include tasks that allow for technology as a tool, such as graphing calculators? Do we provide tasks that assess appropriate use of tools or modeling?

Note: Your end-of-unit assessments should not be of either extreme—all multiple-choice or all constructed-response questions.

4. Are scoring rubrics clear and appropriate?

Discuss: Are the scoring rubrics to be used for every task clearly stated on the assessment? Do our scoring rubrics (total points or proficiency scale) make sense based on the complexity of reasoning for the task? Are the scoring points or scale assigned to each task appropriate and agreed upon by each teacher on our team? Is there clear understanding about the student work necessary to receive full credit for each assessment task or question? Is it clear to each team member how partial credit will be assigned?

Figure 2.2: Team discussion tool—High-quality assessment evaluation.

continued →

5. Are the directions clear?

Discuss: What does clarity mean to each member of our team? Are any of the directions we provide for the different assessment tasks confusing to the student? Why?

Note: The verbs (action words) you use in the directions for each set of tasks or problems are very important to notice when discussing clarity. Also, be sure that in the directions you clearly state the student work you expect to see and will grade using points or a scale.

6. Is the academic language precise and accessible?

Discuss: Are the vocabulary and notation for each task we use on our common assessment clear, accessible, and direct for students? Do we attend to the precision of language used during the unit, and do students understand the language we use on the assessment?

Note: Does the assessment instrument include the proper language supports for all students?

7. Does the visual presentation provide space for student work?

Discuss: Do our students have plenty of space to write out solution pathways, show their work, and explain their thinking for each task on the assessment instrument?

Note: This criterion often is one of the reasons not to use the written tests that come with your textbook series. You can use questions from the test bank aligned to your instruction, but space mathematics tasks and assessment questions as needed to allow plenty of room for students to demonstrate their understanding.

8. Do we allot enough time for students to complete the assessment?

Discuss: Can our students complete this assessment in the time allowed? What will be our procedure if they cannot complete the assessment within the allotted time so all students receive equitable opportunities for demonstrating learning?

Note: Each teacher on the team should complete a full solution key for the assessment he or she expects of students. For upper-level students, it works well to use a time ratio of three to one (or four to one) for student to teacher completion time to estimate how long it will take students to complete an assessment. For elementary students, it may take much longer to complete the assessment. All teachers should use the agreed-on time allotment.

Visit go.SolutionTree.com/MathematicsatWork for a free reproducible version of this figure.

You should expect to eventually write common assessment tests that would score fours in all eight categories of the assessment evaluation.

For deeper clarification and explanation on each of the evaluation criteria in the assessment tool, use figure 2.2 (page 15) to reflect on how your team rates and scores on your common assessment instruments.

TEACHER *Reflection*

Based on the criteria from figures 2.1 and 2.2, summarize the result of your evaluation of an assessment you created and used with your students.

What action can you immediately take to improve your assessments?

_____

_____

_____

_____

_____

_____

_____

_____

The first four design criteria of the assessment instrument evaluation tool (identification and emphasis on essential learning standards, balance of higher- and lower-level cognitive demands, variety of assessment-task formats and use of technology, and appropriate scoring rubric) are perhaps the most important, and yet, also the most limiting aspect of many mathematics unit assessments—both during and at the end of a unit. They are vital to making your assessments "work" for students, as Steve Jobs indicates in the opening quote for this chapter.

These first four mathematics assessment design criteria can often create places of *great inequity* in your mathematics assessment process and professional work. For that reason, we provide extended details for each of these four criteria to help you and your colleagues' growth and development in designing your common mathematics assessments.

TEAM RECOMMENDATION

Design High-Quality Common Unit Assessments

Use figures 2.1 and 2.2 to determine the current strengths and areas to improve on your team common assessments.

- Commit to creating your common unit assessments before the unit begins.

- Respond to the prompt: What is your team plan to improve your common unit assessments this year?

Identification and Emphasis on Essential Learning Standards

Your team can organize assessments in many ways, such as the following.

- Format (multiple choice first and constructed response second)

- Difficulty (from easier tasks to more difficult tasks)

- Order (place the tasks in the order they were identified for the assessment)

- Essential learning standard

The essential learning standards should be the *organizational driver for your assessment design* because the essential learning standards sit at the top of the instructional framework "to signify that setting goals is the starting point for all decision making" (Smith, Steele, & Raith, 2017, p. 195).

Your team should list an essential learning standard for the unit, followed by the mathematics tasks or questions you believe represent evidence of student learning and proficiency for that standard. There should be no more than three to six such standards on any end-of-unit mathematics assessment and perhaps one to two on a unit quiz or mid-unit assessment.

There are many names for the types of standards your team will use, such as *power standards*, *priority standards, promise standards,* and so on. When your team works within a PLC culture and for the purposes in this book, we suggest using the term *essential learning standards* for the unit. These are the *big idea* standards you expect your students to learn.

TEACHER *Reflection*

Review the criteria from figures 2.1 and 2.2. *How should you organize your assessments?* This is a good first question for your team to ask. Describe your current practice.

_____

_____

_____

_____

_____

_____

Rick DuFour and Bob Marzano (2011) ask you to think of the essential learning standards as part of your team's guaranteed and viable curriculum for each unit. A *guaranteed curriculum* is one that promises the community, staff, and students that a student in school will learn specific content and processes regardless of the teacher he or she receives. A *viable curriculum* refers to ensuring there is adequate time for all students to learn the guaranteed curriculum during the school year.

Clear team agreement on the essential learning standards for the unit is a priority. It is the starting point for answering PLC critical question 1 (DuFour et al., 2016): What do we want all students to know and be able to do? This is the launching point for more equitable learning experiences for every student in your grade level or course.

The three to six essential learning standards also inform your team's intervention design and response to critical questions 3 and 4, discussed in part 2 of this book (page 65).

Ultimately, the reason for designing your common assessments around the organizational driver of the essential learning sstandards is to identify where to target future emphasis for students, including:

1. Mathematics tasks or questions, such as the proper distribution and ensured standards alignment of those mathematical tasks placed on the common assessment

2. Team-designed student intervention and extension activities

3. Instructional decisions during and after a unit (We discuss this more thoroughly in *Mathematics Instruction and Tasks in a PLC at Work* in this series.)

Your collaborative team may want to use or create a proficiency map that identifies, by unit, essential learning standards with which students should be proficient by the end of each unit. A quick overview of unit pacing can help your team make connections between standards and what students should learn, which in turn informs quality assessment and a planned team response to student learning. (Visit **go.SolutionTree .com/MathematicsatWork** for examples of several K–5 unit-by-unit, grade-level proficiency maps.)

TEACHER *Reflection*

Look closely at your students' current assessments. How many essential learning standards are you listing for the assessments? Too many? Too few? If you do not list the standards, can you identify how many the tests assess?

_____

_____

_____

_____

_____

_____

Be sure your mathematics assessments, unit tests, and quizzes do not list too many essential learning standards. In order to design high-quality mathematics assessments your team and students can use for formative purposes, focus only on the big idea and most essential learning standards. Also, be sure to write the essential learning standards in student friendly-language using *I can* statements. This helps students verbalize their part in the assessment, learning, and reflection process.

However, the *big idea* standards generally do not provide enough detail for your daily lessons. You most likely use or have heard of many terms for your lesson focus, such as *learning objective, learning target, daily objective, daily learning standard, lesson objective*, and so on. These more detailed and specific standards for the daily lesson help you and your students unwrap the essential learning standards (the big ideas for the unit) into daily skills and concepts to use for the assessment tasks (questions) on the test. For the purposes of our PLC work, we reference these specific parts of standards as *learning targets*.

Your daily learning targets help support the tasks selected for your common assessments but are too specific for test organization purposes. The essential learning standards in each unit comprise sections of the assessment and allow for student reflection and continued learning. (See pages 20–21 for specific student reflection examples.)

To help you compare and contrast the difference between essential learning standards and daily learning targets, figure 2.3 (page 20) provides a sample that illustrates a grade 4 fractions unit. The essential learning standards it lists in the left column present a more formal representation of each essential learning standard, similar to how it might appear in your state standards.

The middle column shows essential learning standards as a teacher would write them on the assessment or test in student-friendly *I can* language. The right column unwraps each standard into daily learning targets for your lesson design and planning using combinations of verbs and noun phrases in the original formal standard.

Using the sample grade 4 fractions unit, your team could create shorter common assessments for each of the essential learning standards throughout the unit. You and your students could then use the results to proactively receive feedback for additional instruction and help *before* the end-of-unit assessment. (Visit **go.Solution Tree.com/MathematicsatWork** to find similar models for first grade, seventh grade, and high school.)

TEACHER *Reflection*

How does your team create and develop its understanding of the essential learning standards for the unit?

How do you break down the essential learning standards for assessment purposes into daily learning targets for instruction purposes?

_____

_____

_____

_____

_____

When you work with your colleagues to determine the essential learning standards for common unit assessments, you ensure equitable learning expectations and experiences for students in classrooms across your team.

TEAM RECOMMENDATION

Determine Essential Learning Standards for Student Assessment and Reflection

- Identify the formal learning standards students must be proficient with by the end of the unit.

- Group the standards as needed to create three to six essential learning standards for the unit. Write these using student-friendly language and begin each with *I can* statements.

- As a team, make sense of the essential learning standards and discuss what students must know and be able to do to demonstrate proficiency.

Formal Unit Standards (Generic state standard language)	Essential Learning Standards for Assessment and Reflection (Uses student-friendly language)	Daily Learning Targets (Explains what students should know and be able to do; unwrapped standards)
1. Explain why a fraction $\frac{a}{b}$ is equivalent to a fraction $\frac{(n \times a)}{(n \times b)}$ by using visual fraction models, with attention to how the number and size of the parts differ, even though the two fractions themselves are the same size. Use this principle to recognize and generate equivalent fractions.	I can explain why fractions are equivalent and create equivalent fractions.	• Explain why a fraction $\frac{a}{b}$ is equivalent to a fraction $\frac{(n \times a)}{(n \times b)}$ by: ‣ Using visual fraction models ‣ Telling how the number and size of the parts differ even though the fractions are the same size • Generate equivalent fractions.
2. Compare two fractions with different numerators and different denominators, for instance, by creating common denominators or numerators or by comparing to a benchmark fraction such as $\frac{1}{2}$. Recognize that comparisons are valid only when the two fractions refer to the same whole. Record the results of comparisons with symbols >, =, or <, and justify the conclusions.	I can compare two fractions and explain my thinking.	• Compare two fractions with different numerators and different denominators by: ‣ Creating common numerators ‣ Creating common denominators ‣ Comparing to a benchmark fraction • Record fraction comparisons using <, =, or >. • Justify fraction comparisons (for instance, using visual models). • Recognize that comparisons are valid only when the two fractions refer to the same whole.
3. Understand a fraction $\frac{a}{b}$ with a > 1 as a sum of fractions $\frac{1}{b}$. a. Understand addition and subtraction of fractions as joining and separating parts referring to the same whole. b. Decompose a fraction into a sum of fractions with the same denominator in more than one way, recording each decomposition by an equation. Justify decompositions, e.g., by using a visual fraction model. c. Add and subtract mixed numbers with like denominators (for example, by replacing each mixed number with an equivalent fraction, and/or by using properties of operations and the relationship between addition and subtraction).	I can add and subtract fractions and show my thinking.	• Compose a fraction using unit fractions. • Add fractions by joining parts of the same whole. • Subtract fractions by separating parts of the same whole. • Decompose a fraction more than one way into a sum of fractions with like denominators. • Add mixed numbers with like denominators using: ‣ Equivalent fractions ‣ Properties of operations ‣ Relationship between addition and subtraction • Subtract mixed numbers with like denominators using: ‣ Equivalent fractions ‣ Properties of operations ‣ Relationship between addition and subtraction

Formal Unit Standards (Generic state standard language)	Essential Learning Standards for Assessment and Reflection (Uses student-friendly language)	Daily Learning Targets (Explains what students should know and be able to do; unwrapped standards)

4. Apply and extend previous understandings of multiplication to multiply a fraction by a whole number.
 a. Understand a fraction $\frac{a}{b}$ as a multiple of $\frac{1}{b}$. For example, use a visual fraction model to represent $\frac{5}{4}$ as the product $5 \times (\frac{1}{4})$, recording the conclusion by the equation $\frac{5}{4} = 5 \times (\frac{1}{4})$.
 b. Understand a multiple of $\frac{a}{b}$ as a multiple of $\frac{1}{b}$, and use this understanding to multiply a fraction by a whole number (for example, use a visual fraction model to express $3 \times (\frac{2}{5})$ as $6 \times (\frac{1}{5})$, recognizing this product as $\frac{6}{5}$. (In general, $n \times (\frac{a}{b}) = \frac{(n \times a)}{b}$.)
	I can multiply a fraction by a whole number and explain my thinking.	• Multiply a fraction by a whole number.
		 • Show that a fraction $\frac{a}{b}$ is a multiple of $\frac{1}{b}$ and write an equation to match.
		 • Multiply a fraction by a whole number using the idea that $\frac{a}{b}$ is a multiple of $\frac{1}{b}$.
		   ‣ Use a visual model.
		   ‣ Write expressions.

3d. Solve word problems involving addition and subtraction of fractions referring to the same whole and having like denominators (e.g., by using visual fraction models and equations to represent the problem).	I can solve word problems involving fractions.	• Solve word problems involving addition and subtraction of fractions with like denominators referring to the same whole using:
		   ‣ Visual fraction models
		   ‣ Equations
		 • Solve word problems involving multiplication of a fraction by a whole number.
		   ‣ Use a visual model.
		   ‣ Write equations.

4c. Solve word problems involving multiplication of a fraction by a whole number (for example, by using visual fraction models and equations to represent the problem). For example, if each person at a party will eat $\frac{3}{8}$ of a pound of roast beef, and there will be 5 people at the party, how many pounds of roast beef will the party need? Between what two whole numbers does your answer lie?		

Figure 2.3: Sample grade 4 fractions unit—Essential learning standards.

Visit go.SolutionTree.com/MathematicsatWork for a free reproducible version of this figure.

Once your team begins the common assessment design (your high-quality surfboard) by building it around the essential learning standards for the unit, what's next for building your unit assessments? The next step is revealed in your choices for the mathematical tasks you believe align to and represent each essential learning standard on the assessment. After you determine the essential learning standard, your *choice of tasks drives everything you do* because "tasks are the vehicles that move students from their current understanding toward the essential learning standard" (Smith et al., 2017).

Choosing mathematical tasks is one of your most important professional responsibilities as a mathematics teacher. It permeates everything you do. You decide each day the mathematical tasks students perform and ask in class. You then decide the tasks and questions to assign for homework. You also work with your team to decide the nature of the mathematical tasks you place on the exam, as well as the ones to use for intervention on standards.

That is a lot of power.

And, it is why working on these decisions *together* as a team is so important. It is why the second criterion in figure 2.1 (page 14) is the rigor revealed by the cognitive demand of the mathematical tasks you have chosen for the assessment.

Balance of Higher- and Lower-Level-Cognitive-Demand Mathematical Tasks

There are several ways to label the cognitive demand or rigor of a mathematical task; however, for the purpose of this book, mathematical tasks are classified as either lower-level cognitive demand or higher-level cognitive demand, as Margaret S. Smith and Mary Kay Stein (1998) define in their task analysis guide printed in full in the appendix: "Cognitive-Demand-Level Task Analysis Guide" (page 111).

Lower-level-cognitive-demand tasks typically focus on memorization by performing standard or rote procedures without attention to the properties that support those procedures (Smith & Stein, 2011). *Higher-level-cognitive-demand tasks* are those for which students do not have a set of predetermined procedures to follow to reach resolution or, if the tasks involve procedures, they

require that students justify why and how to perform the procedures.

Figure 2.4 (page 23) shows examples of lower-level- and higher-level-cognitive-demand tasks (either for use in class or on an assessment) for various grade levels. Use the most appropriate grade-level question, and discuss why each question meets the cognitive-demand levels using the descriptions that the "Cognitive-Demand-Level Task Analysis Guide" provides.

TEACHER *Reflection*

After using figure 2.4 and the appendix (page 111) to explain why the grade-level tasks were low level or high level, think about a concept that students in your grade level must learn and write or describe a low-level and high-level-cognitive-demand task for that concept.

Describe how your cognitive-demand choices for each task reveal underlying student understanding of an essential learning standard.

_____

_____

_____

_____

_____

_____

_____

_____

_____

A key word regarding rigor is *balance*, as noted in criterion 3 in figure 2.2 (page 15). Your common assessments should reveal a balance of procedural fluency and conceptual understanding proficiency on all during and end-of-unit assessments. This student proficiency will be revealed through the mathematical tasks you choose to place on the assessment, including tasks that have an application orientation. Generally, the rigor-balance ratio should be about 30/70 (lower- to higher-level cognitive demand) on an end-of-unit assessment.

Directions: Choose the most appropriate grade level that follows and discuss why each of the questions meets the cognitive-demand levels to which it's assigned using the descriptions for lower- and higher-level-cognitive-demand tasks in the appendix (page 111).

Grade 1: I can compare two numbers and write my answer using <, =, or >.

Lower-level-cognitive-demand task:	**Higher-level-cognitive-demand task:**
Write <, =, or > in the circle to compare the numbers.	Explain why 45 < 62. Use numbers, pictures, or words.
23 $\bigcirc$ 51	

Grade 4: I can add and subtract fractions and show my thinking.

Lower-level-cognitive-demand task:	**Higher-level-cognitive-demand task:**
Add and show how you know your answer is correct.	Martin runs $\frac{5}{6}$ of a mile and then walks $2\frac{2}{6}$ mile. How much farther does he walk than run? Show how you know your answer is correct.
$\frac{5}{6} + 2\frac{3}{6} = \boxed{}$	

Grade 7: I can solve multistep ratio and percent problems.

Lower-level-cognitive-demand task:	**Higher-level-cognitive-demand task:**
What is a $40 shirt worth if it is on sale for 20% off the original price?	Carla wants to buy tennis shoes that have a price tag of $75. They are on sale for 20% off the original price, and she has a coupon to use that reduces the price an additional 10% off the sale price. Will Carla buy the shoes for 30% off the original price? Explain why or why not.

Figure 2.4: Examples of lower- and higher-level-cognitive-demand tasks.

continued →

High school (algebra 1): I can graph linear functions and interpret their key features.

Lower-level-cognitive-demand task:

The following table shows the height of a plant in centimeters for each day of growth for ten days. Graph the points. Explain the meaning of the slope of the graph.

Day	Plant Height (cm)
0	0.3
2	1.1
4	1.9
7	3.1
10	4.3

Higher-level-cognitive-demand task:

Samantha collects cans for recycling in Oregon. She makes a graph to show the money she could earn—y, for recycling x cans. Two points on her graph are (10, 1.50) and (25, 3.75).

a. Draw a graph to show the amount of money Samantha earns for a given number of recycled cans.

b. Explain the domain of the function.

c. In California, Samantha's graph would look different. It would pass through the points (6, 0.60) and (18, 1.80). How does the money she could earn in each state compare? Use key features of the graph to explain your comparison.

TEACHER *Reflection*

Think about your current end-of-unit assessments. In general, describe the balance of lower-level-cognitive-demand questions to higher-level-cognitive-demand questions on your assessments.

_____

_____

_____

_____

_____

_____

_____

TEAM RECOMMENDATION

Agree on the Lower-Level and Higher-Level-Cognitive-Demand Mathematical Tasks Chosen for Common Unit Assessments

- As a team, identify examples of lower- and higher-level-cognitive-demand tasks for the essential learning standards in a unit.

- Determine which tasks to place on the assessment and use a 30/70 ratio of low- to high-level tasks as a guide for ensuring a balance of rigor on the assessment.

Your team should not be expected to create each and every mathematics assessment task used on the unit assessment. You need to determine the sources you will use to choose the mathematical tasks placed on the assessment for each essential learning standard. Your team may decide to start with a publisher assessment or previously used assessment to gather quality problems, questions, and tasks. From there, you may search websites for assessment questions and tasks using some of the online resources for mathematics assessment support (visit **go.SolutionTree.com/MathematicsatWork** for the free reproducible "Online Resources Reference Guide for Mathematics Support"), or access tasks in textbooks or other reliable resources. Your team may even create some of your own assessment questions to gather evidence of student learning as you build assessments together.

If you are a singleton teacher, and you are teaching a grade level or course with released state assessment questions or nationally released assessment questions, you should consider placing some of these tasks on your assessments to compare the responses of your students to those of comparable students in your state or across the country.

When you work with your colleagues to agree on the lower-level- and higher-level-cognitive-demand tasks for your assessment aligned to the essential learning standards, you create improved equity in student learning expectations and rigor and erase the inequities that the individual interpretation of rigor from teacher to teacher on your team causes.

No matter where you resource your assessment questions and tasks, you most likely will encounter many different formats of questions with and without various forms of technology as well. This then leads to the third criteria for you and your team to consider as you build a high-quality assessment: *variety of assessment-task formats and use of technology.*

TEACHER *Reflection*

Currently, what is your favorite format to use for the mathematics questions or tasks on your assessments? Do you prefer multiple choice, open ended, true or false, or matching or those involving manipulatives, technology, online material, or paper and pencil?

Why do you have these format preferences?

_____

_____

_____

_____

_____

_____

_____

_____

_____

These choices, when made in isolation from your colleagues, can also become a source of inequity in your assessment of students. Discussion of these formatting choices is next.

Variety of Assessment-Task Formats and Technology

Assessment formats are the types of tasks, tools, and technology students use when taking a common mathematics assessment. When designing most common end-of-unit assessments or common formative assessments for your team to use during a unit, your mathematics assessment tasks will either include selected-response tasks or constructed-response tasks.

- *Selected-response tasks* refer to multiple choice, fill-in-the-blank, matching, true or false, or questions with an immediate right or wrong answer.

- *Constructed-response tasks* refer to a short answer in which students show their reasoning or solution pathway as well as their final answer. In kindergarten and early first grade, this would also include short answer in oral or demonstration form instead of a written expectation. Your team should determine if you will administer the mathematics assessment one on one, in small groups, or with a whole group. Written performance assessment tasks or projects are in the family of constructed-response tasks, as students are showing learning, in real time, which is often complex in reasoning.

Ideally, your common assessments provide students with practice using these various assessment formats.

Selected Response

Often, the most difficult selected-response task formats to design and use accurately are multiple choice and matching. Guidelines are provided to help you evaluate the quality for these two formats.

Multiple Choice

Consider the following five guidelines when using multiple-choice tasks.

1. Be sure every distractor (incorrect answer) results from a common misconception or error students often use for that concept.

2. Be sure each incorrect answer provides your team as much information about student learning as a correct response. If your team cannot think of plausible distractors, consider leaving the question open ended this year and gathering common misconceptions as distractors for next year's assessment.

3. Consider allowing students to identify more than one correct answer, which more easily allows for higher-level-cognitive-demand tasks. Be sure to let students know that more than one answer may be correct in the directions or the stem (question).

4. Consider parallel construction in the responses. If, for example, one distractor or answer is written as a decimal while the rest are written as whole numbers, students often choose the one that looks different from the rest, whether correct or incorrect. The answer and distractors should be similar in length whether written as sentences or number values.

5. Be sure to allow for partial credit (students can show work) when it is time to score multiple-choice assessment questions.

Matching

Consider the following two guidelines when using matching tasks.

1. Make the list students match *from* shorter than the list they match *to*. This prevents faulty data. When the lists are the same length, too many students simply connect the last remaining parts when they get to the last item. If they have made an error previously, they now have made at least two errors in reasoning. To minimize this, if the list they are matching *to* is longer, students are more apt to look at the remaining options and correct their thinking as needed.

2. Allow students to match to any possible response more than one time when matching, again to minimize flawed results from the assessment.

Constructed Response

Consider the following three guidelines when using constructed-response tasks.

1. Be sure the directions for the task explain the work you require of the students. For example, if you and your colleagues expect a student to draw a picture or graph and that is part of your scoring

agreement, you must tell students that they need to include a picture or graph in their solution.

2. Keep the task's wording simple except for academic vocabulary the task requires.

3. Write tasks using familiar contexts for students. For example, avoid a skiing task if students live in an area without easy access to skiing.

TEACHER *Reflection*

Which of the guidelines for selected-response tasks of multiple choice and matching or constructed-response tasks do you already consider when creating your unit assessments?

Which guidelines will your team need to remember when choosing or creating assessment tasks to make your common unit assessments stronger?

_____

_____

_____

_____

_____

_____

_____

Manipulatives and Technology

Consider the following three guidelines when using manipulatives or technology as part of the common mathematics assessment.

1. Find team agreements on how you want to use technology as a tool. When should students be able to have the technology as a choice to select from when solving the mathematical tasks?

2. To build conceptual knowledge, students need opportunities to interact with mathematics and digital tools on the assessment. Can you provide access to the technology without lowering the cognitive demand or creating a crutch in learning?

3. It is best to use manipulatives and technology tools on tasks you design to assess conceptual

understanding and application or the modeling type of mathematics tasks.

Since middle school can also use embedded technology on state assessments as early as sixth grade, and when taking the SAT and ACT exams as high school students, it wise to address the technology issue as part of your local common assessment tools as well. When students use technology tools and manipulatives during assessment, they can learn through exploration and streamline computations focusing on more complex mathematical skills.

However, be sure not to create a team inequity. If you plan to use manipulatives or technology in your common assessment instruments, all the team members should also allow students to benefit from using the same manipulatives or specific technology as part of the student instruction and assessment experience.

TEACHER *Reflection*

To which manipulatives, technology, or other tools do students need access to demonstrate proficiency on your common unit assessments? How will your team provide equal access to these tools as part of mathematics instruction and assessment?

_____

_____

_____

_____

_____

_____

_____

For example, if you are assessing sorting and classifying skills in kindergarten, be sure each team member uses the same manipulatives (colored bears or pattern blocks) to create equitable and meaningful assessment results for all students. Similarly, if you are using fraction tiles in the intermediate grades, ensure all students have access to or are told to use the same fraction tiles. At the secondary level, access to and the required use of graphing calculators or algebra tiles should be consistent across all classrooms and for all students.

TEAM RECOMMENDATION

Make a Plan for Assessment Methods and Appropriate Tools

- Determine which assessment tasks should be selected response and constructed response.

- Examine the guidelines for writing quality selected-response and constructed-response tasks.

- Reach agreement on the answer to these questions: When should our students choose and use tools or technology on a common unit assessment? How will all students have equal access to the tools and technology?

TEACHER *Reflection*

Think about your current assessments.

If you score assessments using a points-per-question scale or rubric, do you clearly state on the test the points possible for each task or mathematics problem? Do your total points for the assessment make sense based on the task's complexity of reasoning required for successful student demonstration of work?

If you use a standards-based grading rubric, does each team member agree on the meaning of 1, 2, 3, or 4 as a grade for each assessment task or group of questions? How do you determine a student's level of proficiency for each essential learning standard? Does the scaled score make sense based on the task's complexity of reasoning?

_____

_____

_____

_____

_____

_____

_____

_____

_____

_____

_____

_____

This now brings you to the fourth criteria for writing high-quality common assessments: appropriate and clear scoring rubrics.

Appropriate and Clear Scoring Rubrics

At this point, your team has created a common mathematics assessment based on targets that align to the essential learning standards for the units. Your team has also determined the best and cognitively balanced mathematical tasks needed to gather evidence of student learning. How then, should you grade, score, and evaluate each assessment task? Reaching agreement on this issue is a primary responsibility of your professional work together.

Scoring an assessment task is a two-step process.

1. The team reaches agreement on the *scoring value* of each assessment task.

2. The team reaches agreement on the evidence of understanding requirements for a student to receive full or partial credit for a response to each mathematical task.

Review the assessment questions in figure 2.5, and select a grade level most appropriate to your current teaching assignment. First, determine how many points you would score for the task or how you would describe

proficiency levels 1–4 for the task. If using a proficiency scale, 1 represents minimal understanding, 2 represents partial understanding, 3 represents proficient understanding, and 4 represents exceeding grade-level understanding or expectations.

Then, as a team, use the team discussion tool in figure 2.6 (page 30) to calibrate your scoring decisions with your colleagues. (Visit **go.SolutionTree.com /MathematicsatWork** to find additional grade-level tasks online.)

Directions: Choose a grade-level task most appropriate to your current teaching assignment. Determine how many points you would use to score the task or describe how you would give a proficiency-based score of 1–4 for the task.

Kindergarten Task

Essential learning standard: I can solve addition and subtraction word problems within 10 and show my thinking.

Sam has 8 toy cars. He gives 3 toy cars to his brother. How many toy cars does Sam have now? Show how you know your answer is correct.

Grade 3 Task

Essential learning standard: I can represent and solve two-step word problems using the four operations.

Jo buys 5 small baskets of strawberries. Each basket holds 9 fresh strawberries. She uses 15 strawberries for a dessert. How many strawberries does Jo have left over? Use words, numbers, pictures, or a combination of these to show how you know your answer is correct.

Grade 6 Task

Essential learning standard: I can solve word problems by writing equations and solving them. (*Equations are of the form $x + p = q$ and $px = q$ for cases in which p, q, and x are all nonnegative rational numbers.)

Joshua makes a dessert. He uses 5 equal scoops of sugar and uses a total of $3\frac{1}{2}$ cups of sugar. How much sugar is in each scoop? Use words, numbers, pictures, or a combination of these to show how you know your answer is correct.

High School Task

Essential learning standard: I can write equations or inequalities for constraints to a modeling problem and interpret possible solutions.

A club is selling T-shirts and sweatshirts for a fundraiser. Its budget is $1,800, and the club wants to order at least 250 items. The club must buy at least as many T-shirts as they buy sweatshirts. Each T-shirt costs $6, and each sweatshirt costs $12.
- Write a system of inequalities to represent the situation.
- Graph the inequalities.
- If the club buys 150 T-shirts and 150 sweatshirts, will the conditions be satisfied?
- What is the maximum number of sweatshirts the club can buy and still meet the conditions?

Figure 2.5: Team discussion tool—Sample assessment questions for team scoring.

Visit go.SolutionTree.com/MathematicsatWork for a free reproducible version of this figure.

How did you and your teammates determine the scoring of the mathematical task for your selected grade-level problem? Did students receive one point for their reasoning and one point for the correct answer, or no credit if the answer is incorrect? Is each part of the question worth one point, two points, four points, or more? What must a student show to earn a score of 1, 2, 3, or 4 if you used a proficiency scale? If you give an assessment in real time orally how do you document the score in real time?

If your collaborative team does not engage in conversations about scoring, inequities will persist across your team, and ultimately, assigned grades will be inaccurate from teacher to teacher. Your team needs to articulate and agree on the scoring for all exam tasks and questions before asking students to take the test.

What might it look like to write scoring agreements for common assessment tasks? Figure 2.7 (page 31) shows possible scoring rubrics for the tasks in figure 2.5. It assigns points for showing reasoning as well as the final answer.

Collaborative-Team-Task-Scoring Discussion Prompts

Directions: With your collaborative team, examine the following questions for the task you chose from figure 2.5 (page 29).

1. How would you assign the points for different parts of the solution or describe levels of proficiency using a scale of 1–4?

2. If the point value for the task is greater than 1, how could a student earn partial credit? If you use a proficiency scale, how does a student earn a 2?

3. What evidence of student learning is required to receive full credit?

4. How will the team ensure the scoring of the task will be consistent between all team members?

5. How is the expected scoring rubric for the assessment task articulated to students?

Figure 2.6: Team discussion tool—Scoring assessment tasks.

*Visit **go.SolutionTree.com/MathematicsatWork** for a free reproducible version of this figure.*

Directions: Analyze the points or proficiency scale scores assigned to each task. Do you and your colleagues agree or disagree? Why?

Kindergarten Task

Essential learning standard: I can solve addition and subtraction word problems within 10 and show my thinking.

Sam has 8 toy cars. He gives 3 toy cars to his brother. How many toy cars does Sam have now? Show how you know your answer is correct.

Points	Proficiency Scale
Two points total: • 1 point for equation or work showing $8 - 3 = ?$ or $3 + ? = 8$ • 1 point for answer (5 toy cars) with or without units	4—Student solves the task correctly with work to show the answer is correct and checks the work or shows two ways to justify the answer, including toys, cars, or toy cars as units at the end of the answer. 3—Student solves the task correctly with work to support the answer and may have reversal of numbers or not have the units as part of the answer. 2—Student has correct work with an incorrect answer or a correct answer, but the work does not support the answer. 1—Student shows minimal understanding in the work and the answer. (Student might solve $8 + 3 = ?$ correctly.)

Grade 3 Task

Essential learning standard: I can represent and solve two-step word problems using the four operations.

Jo buys 5 small baskets of strawberries. Each basket holds 9 fresh strawberries. She uses 15 strawberries for a dessert. How many strawberries does Jo have left over? Use words, numbers, pictures, or a combination of these to show how you know your answer is correct.

Points	Proficiency Scale
Three points total: • 1 point for finding the total number of strawberries ($5 \times 9 = 45$) • 1 point for finding the strawberries left (for example, $45 - 15 = 30$) • 1 point for answer (30 strawberries)	4—Student solves the task correctly with work to show the answer is correct and checks the work or shows two ways to justify the answer, including strawberries as units at the end of the answer. 3—Student solves the task correctly with work to support the answer and may not have the units as part of the answer. 2—Student has correct work with an incorrect answer or a correct answer, but the work does not support the answer. 1—Student shows minimal understanding in work and answer. (Student might use the numbers 5, 9, and 15 with random operations and perform the calculations correctly.)

Grade 6 Task

Essential learning standard: I can solve word problems by writing equations and solving them.
(Equations are of the form $x + p = q$ and $px = q$ for cases in which p, q, and x are all nonnegative rational numbers.)

Joshua makes a dessert. He uses 5 equal scoops of sugar and a total of $3\frac{1}{2}$ cups of sugar. How much sugar is in each scoop? Use words, numbers, pictures, or a combination of these to show how you know your answer is correct.

Points	Proficiency Scale
Three points total: • 1 point for knowing how to solve $3\frac{1}{2} \div 5 = ?$ or $5 \times ? = 3\frac{1}{2}$ • 1 point for performing the calculation correctly or using models correctly • 1 point for answer ($\frac{7}{10}$ cup)	4—Student solves the task correctly with work to show the answer is correct and checks the work or shows two ways to justify the answer, including cup, of a cup, or cups as units at the end of the answer. 3—Student solves the task correctly with work to support the answer and may not have the units as part of the answer. 2—Student has an error in the solution pathway though the answer matches the student work, or a student has a correct answer, but the work does not support the answer. 1—Student shows minimal understanding in the work and the answer. (Student might use another operation with the numbers $3\frac{1}{2}$ and 5.)

Figure 2.7: Example of task points or proficiency scale scores for tasks in figure 2.5.   continued →

High School Task

Essential learning standard: I can write equations or inequalities for constraints to a modeling problem and interpret possible solutions.

A club is selling T-shirts and sweatshirts for a fundraiser. Its budget is $1,800, and the club wants to order at least 250 items. The club must buy at least as many T-shirts as it buys sweatshirts. Each T-shirt costs $6, and each sweatshirt costs $12.

    a. Write a system of inequalities to represent the situation.
    b. Graph the inequalities.
    c. If the club buys 150 T-shirts and 150 sweatshirts, will the conditions be satisfied?
    d. What is the maximum number of sweatshirts the club can buy and still meet the conditions?

Points	Proficiency Scale
Nine points total: a. 3 points—one for each of the inequalities $t + s \geq 250$ $6t + 12s \leq 1,800$ $t \geq s$ Student may include $t \geq 0$ and $s \geq 0$. b. 3 points for graphing the feasible region correctly 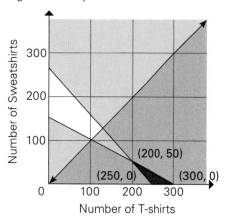 c. 1 point for answer—no (ideally with justification from graph or equations) d. 2 points for answer—solving intersection of equations and interpreting the solution: 50 sweatshirts	4—Student solves the task correctly with work to show the answers are correct and verifies the work for part d or shows two ways to justify the answer for part d. 3—Student solves the task correctly with work to support the answers or student has a simple mistake (including an incorrect inequality in part a) in the solution and then builds incorrect answers, but they are correct in the work due to the minor error. Student might not include $t \geq 0$ and $s \geq 0$, but it is shown in the graph. 2—Student has a conceptual error in the solution pathway, although the answer matches the student work or a correct answer, but the work does not support the answer. 1—Student shows minimal understanding in the work and the answer (for example, only uses one inequality or writes everything as equations and is not sure how to graph, and so on).

TEACHER *Reflection*

How similar or different were your ideas for scoring the task you chose in figure 2.5 (page 29), from the other teachers on your team?

Why is it important to reach common scoring agreements for the tasks on your assessments?

_____

_____

_____

_____

Earning a 4 on a proficiency scale is difficult for a single task, but it offers ideas for student work that exceed grade-level expectations for proficiency. (You can find more information about proficiency scales on page 59.) How similar or different are your team-scoring agreements for the task you chose?

Having strong norms in place will help your team learn how to agree on the scoring rubric for your common assessments. Initially, you might need a facilitator whether it is an instructional coach or other teacher or support person. We provide detailed information on how to reach these agreements in *Mathematics*

Coaching and Collaboration in a PLC at Work. (Visit **go.SolutionTree.com/MathematicsatWork** to find additional sample mathematics tasks for grades K–12.)

Figure 2.8 provides sample student work for each task from figure 2.5. Using the scoring rubric your team agrees on for the task, evaluate and score student work for the task appropriate to your grade level. Consider how your team would score student work on the assessment question based on the complexity of reasoning the student requires. Is the student proficient? How would you know? Once again, you can use the team discussion prompts in figure 2.6 (page 30) to help team conversations around common scoring, using the selected assessment task. It might be interesting to "blind" double score the task with a partner and discover how similar your scores are for the same student work. (We discuss scoring calibration further in part 2, page 65.)

If grading the kindergarten tasks, how do the scores for the two students differ, if at all, knowing student A has the incorrect answer and student B uses the incorrect symbol to show his or her thinking? If grading the third-grade task, how might counting errors influence the student's overall score? If grading the sixth-grade task, how might the check that verified the incorrect answer factor into the overall score, if at all? If grading the high school task, how did the incorrect inequality symbols in part a influence the overall score? How does the student arrive at the answer of fifty sweatshirts?

Directions: Evaluate student work using your team scoring agreements. Is the student proficient? Does everyone on your team agree?

Kindergarten Task

Essential learning standard: I can solve addition and subtraction word problems within 10 and show my thinking.

Sam has 8 toy cars. He gives 3 toy cars to his brother. How many toy cars does Sam have now? Show how you know your answer is correct.

Student work:

Student A	Student B
$8 - 3 = 6$ ⑧ ③ ⑥	☒ ☒ ☒ ☐ ☐ $8 + 3 = 5$ ☐ ☐ ☐ ☐

Grade 3 Task

Essential learning standard: I can represent and solve two-step word problems using the four operations.

Jo buys 5 small baskets of strawberries. Each basket holds 9 fresh strawberries. She uses 15 strawberries for a dessert. How many strawberries does Jo have left over? Use words, numbers, pictures, or a combination of these to show how you know your answer is correct.

Student work:

15 strawberries

$9 + 6 = 15$

(●) + (●) + (●) + (●) + 28 left over

(●) = 54

Figure 2.8: Sample assessment questions with team scoring.

continued →

Grade 6 Task

Essential learning standard: I can solve word problems by writing equations and solving them. (Equations are of the form $x + p = q$ and $px = q$ for cases in which p, q, and x are all nonnegative rational numbers.)

Joshua makes a dessert. He uses 5 equal scoops of sugar and uses a total of $3\frac{1}{2}$ cups of sugar. How much sugar is in each scoop? Use words, numbers, pictures, or some combination of these to show how you know your answer is correct.

Student work:

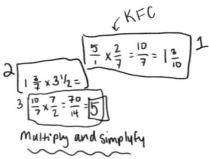

High School Task

Essential learning standard: I can write equations or inequalities for constraints to a modeling problem and interpret possible solutions.

A club is selling T-shirts and sweatshirts for a fundraiser. Its budget is $1,800, and the club wants to order at least 250 items. The club must buy at least as many T-shirts as it buys sweatshirts. Each T-shirt costs $6, and each sweatshirt costs $12.

   a. Write a system of inequalities to represent the situation.
   b. Graph the inequalities.
   c. If the club buys 150 T-shirts and 150 sweatshirts, will conditions be satisfied?
   d. State the maximum number of sweatshirts the club can buy and still meet the conditions.

Student work:

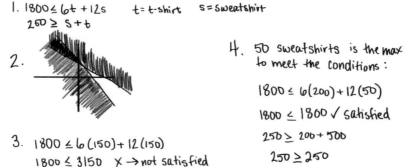

When you score common mathematics assessments privately and, therefore, often differently from one another, inequities can result in feedback and grading reports to students and team member analysis of results.

You have now explored the first four criteria for the assessment instrument quality evaluation tool, which are:

1. Identification and emphasis on essential learning standards (specific feedback to students)

2. Balance of higher- and lower-level-cognitive-demand mathematical tasks

3. Variety of assessment-task formats and use of technology

4. Appropriate and clear scoring rubric (points)

TEACHER *Reflection*

If your team scores with points, how can students earn partial credit for their work?

If your team scores using a proficiency scale, what types of mistakes can a student make and still be considered proficient with tasks? How will your team determine proficiency (level 3)?

_____

_____

_____

_____

_____

TEAM RECOMMENDATION

Determine Common Scoring Agreements for Common Mathematics Assessments

- Determine as a team how you will score assessment tasks on common assessments (points or proficiency scale).

- Clarify how students earn points or how they earn a score of 1–4 on a proficiency scale.

- Determine how students learn how you will evaluate their work on an assessment.

The remaining four criteria address overall formatting and clarity of the common assessment so student results are valid and reliable, including clear directions, academic language, visual presentation, and time allotment. Although they are not of the same high priority as the first four criteria, they are critical to student understanding, and performance during the mathematics assessment.

Clarity of Directions, Academic Language, Visual Presentation, and Time Allotment

In addition to clear essential learning standards, purposeful and balanced mathematics tasks, intentional mathematics assessment task types, and common scoring agreements, reaching team agreements on *how* to format assessment tasks with clear directions and language also impacts student results.

Planning for clarity of directions, academic language, visual presentation, and time allotment means you and your colleagues can have confidence that the format of the assessment is not skewing results, and students can have confidence to persevere when taking the mathematics assessment because it appears and reads as a doable event.

Clarity of Directions

In order for you and your team to know whether or not students are proficient with each essential learning standard, the directions need to clearly specify what they expect students to *do* to fully respond and answer each question. For example:

- For which mathematics tasks must students justify or explain their solutions?

- For which mathematics tasks must students make a table or draw a picture or graph?

- For which mathematics tasks can students use manipulatives or some type of technology?

- For which mathematics tasks must students write expressions or equations or build a model?

The directions clarify how students need to answer specific questions. Pay special attention to the *verbs* you use in the directions as they convey the action students must use to demonstrate understanding.

When you and your colleagues create an answer key for the common assessment and compare your solutions, you will quickly see if any two team members interpreted the directions differently. If different interpretations exist, work as a team to decide if both solutions are acceptable or if you need to clarify the directions further.

Sometimes you will collectively agree that the directions are clear only to find that students interpret them in a way you never anticipated. If this happens, discuss as a team what needs to be revised the next time you give the assessment and how you will score student work this time around.

Academic Language

The vocabulary and symbol notations in mathematics should be precise and comprise a language that allows students to make connections between mathematical concepts and clearly communicate their understanding of those concepts.

TEACHER *Reflection*

How do you currently decide, as a team, if the directions and vocabulary your assessments use are appropriate for and clear to your students?

How do you know if students have enough space to show their work and if they have enough time to finish the assessment?

_____

_____

_____

_____

_____

_____

_____

You and your team model precision when the tasks you place on the assessment use direct, fair, accessible, clearly understood language and consistent symbol notation. In return, you can expect students to demonstrate proficiency with concepts by also using precise academic language in their solutions. You would not, for example, want a student to reference the *top* and *bottom* of a fraction but rather reference the *numerator* or *denominator*. Similarly, you would not want students to write *cos* or *sin* but rather *cos x* or *sin x* and understand the terms read as functions (*cosine of x* and *sine of x*). Precision in language and the solution process greatly influences effective mathematical communication.

Visual Presentation

When creating the assessment, consider how much space you allocate for students to explain their thinking and show their solutions. Is there enough? Too much?

Often publisher assessments do not leave enough space for students to work. If used as is, students might cram their work into the margins or write an answer thinking it is all that is required. Additionally, if tasks are not spaced appropriately and contain a lot of words, students may find the assessment daunting and give up before ever stopping to read and persevere to solve the tasks.

Along with appropriate space for work, visual presentation refers to the exam's neatness and organization. Can students read the assessment and symbol notations clearly? Do you group assessment items by essential learning standard or type (for example, selected-response section, calculator section, and constructed-response section)? Does the organization make sense? Work together to make sure the assessment is visually inviting and clear.

Time Allotment

Time yourself as you complete full solutions to the assessment tasks. Then time how long it takes your students to complete the assessment. Use these timings to "find your number." Your number is what you multiply your time by to better estimate how long it might take your students to complete the assessment. The ratio of time it takes you to complete the assessment to the time it takes your students to complete the assessment may vary by course at the secondary level and by grade level in elementary school.

Compare estimates as a team to determine the duration of time you think it requires for students to take the assessment. How close are you to one another, and how close are you to the actual time it takes students? Do your students receive equitable time expectations from all teachers on your team? If you decide to split the assessment into parts, clarify which parts students will take, in which order, and on which days.

If, however, you find the assessment is shorter or longer than you would like, your team may need to revise it to better match an effective time allotment. Shorter, common, mid-unit assessments should still leave time for instruction in a class period or block of time. Educators often plan common unit assessments for one class period or one block of mathematics time.

Time can also be an issue for kindergarten assessments or mathematics assessments conducted through one-on-one conversations or minigroup sessions, such as at a kidney bean table. To minimize loss of class time, consider how to use written assessments for those essential learning standards from which you will get valid results (for example, writing numbers to twenty can be done as a whole group in kindergarten) and minimize the number of assessments needing one-on-one interactions.

At times, you may need to decide as a team how to conduct one-on-one assessments for students unable to demonstrate their learning through writing, but this does not necessarily need to be the assessment method for all students due to the time structure it requires. It does, however, require a script and agreements across the team to make sure the results come from equitable assessment experiences.

While not as time intensive for your team to address when creating a common assessment as the first four criteria of the assessment instrument quality evaluation tool, directions, academic language, space, and time also influence student results. When directions are unclear, tasks omit important vocabulary or notations, questions are crammed together with limited space for student work, or students don't get enough time to complete the assessment thoroughly, we create inequities in the student learning process. Make sure your team agrees on each of these four issues.

Armed with the criteria for determining the quality of your assessment tools, your team can create common mid-unit assessments to use during the unit and at the end of the unit for continued student learning. We discuss examples of both types of common assessments in the next chapter.

TEAM RECOMMENDATION

Determine Clarity of Directions, Academic Language, Visual Presentation, and Time Allotment

Take the assessment as if you are a student, and then check your responses as a team to ensure the following.

- Directions are clear.

- Academic language (vocabulary and notation) is fair and appropriate.

- Space between tasks is appropriate for the amount of work a student needs to show.

- Time allocated for the assessment is doable for students across the classrooms of your team.

TEACHER *Reflection*

Examine a recent common unit assessment. Read the directions. How clear are they, and how well do they match the common scoring agreements? What verbs do the directions use?

Read the tasks. Which academic language (vocabulary and notation) do they include? What might you need to revise or add to make questions on the assessment clearer for students?

Look at your assessment. Is there space for student work? How crammed are the words on a page? Visually, does the assessment look doable?

How many minutes did it take you to make a complete answer key? How many minutes did it take your students to complete the assessment? What is the teacher-to-student ratio you use when estimating the amount of time that it will take students to complete the assessment?

_____

_____

_____

_____

_____

_____

_____

_____

_____

_____

_____

_____

_____

_____

_____

_____

CHAPTER 3

Sample Common Mathematics Unit Assessments and Collaborative Scoring Agreements

Through frequent assessment, teachers discover when students need additional support to continue their learning.

—*John Hattie, Douglas Fisher, and Nancy Frey*

Every common mathematics unit assessment, no matter how strong it may be, can still be improved—crafted into a stronger tool to more accurately and effectively gather evidence for you and your students as to which essential learning standards students have learned and which they have not learned *yet*. Your continual improvement gathering dependable evidence occurs both during and at the end of assessments.

For some mathematics units or modules, your team may want to create more frequent common assessments to use during a unit. Your team can use the assessment evaluation tool described in chapter 2 to evaluate the quality of these during-the-unit assessments. However, sometimes due to the intent of the essential learning standard, during-the-unit assessments (think quizzes) may not contain a variety of assessment formats and may have a different balance of higher- to lower-level-cognitive-demand tasks.

This chapter provides you with various examples of grade-level unit assessments and scoring rubrics for end-of-unit assessments.

Sample During-the-Unit Common Assessments

Figure 3.1 (page 40) and figure 3.2 (page 41) represent sample mid-unit common assessments for grade 1 and grade 8. Select the sample closest to your grade level and determine if the common assessment will provide the information you need to target additional engagement and support for students still not proficient with the standard once they receive feedback from you on the common mid-unit assessment.

You and your colleagues might also decide to use a common mid-unit assessment exit slip related to a target recently learned. In kindergarten, your common mid-unit assessments might occur at stations or a kidney bean table as you observe student thinking and use a recording tool to document students' answers, rather than use a paper-and-pencil mini-assessment to a target. Other times, you might create a paper-and-pencil assessment along with a script all teachers can follow when administering it.

TEACHER *Reflection*

Describe how you currently use performance assessments and projects based on aligned essential learning standards in order to strengthen your questions and outcomes.

Currently, how do you and your colleagues decide to score the performance task or project?

How do you identify the purpose of the performance task or project and communicate that purpose to create meaning for the student?

_____

_____

_____

Directions: Decide if this common mid-unit assessment provides the information your team needs to determine if students have learned the essential learning standards. Will the results give your team the targeted information it needs to best re-engage students in learning when it gives additional time and support?

Name: _____   Date: _____

Grade 1 Place Value Check
I can compare two-digit numbers.

1. In each box, circle the largest number.   3 points _____

21 19	37 60	42 46

2. Circle all the numbers **less than 40**.   2 points _____

   31   52   16   40   68   25

3. Draw a picture using tens and ones to show why **25 is less than 64**.   2 points _____

4. Write >, =, or < in each circle to compare the numbers.   4 points _____

   a. 35 ◯ 38   c. 12 ◯ 21

   b. 80 ◯ 40   d. 91 ◯ 91

5. Carmen says 53 is less than 39 because 3 ones is less than 9 ones.   2 points _____
    Is Carmen correct? Explain your thinking.

     ☺   ☺?   ☺✋
   I get it!   I get some of it.   I need help.

Figure 3.1: Grade 1 common mid-unit assessment place value sample.

*Visit **go.SolutionTree.com/MathematicsatWork** for a free reproducible version of this figure.*

Directions: Decide if this common mid-unit assessment provides the information your team needs to determine if students have learned the essential learning standards. Will the results give your team the targeted information it needs to best re-engage students in learning when it gives additional time and support?

Name: _____   Date: _____

Grade 8 Solving Check
I can solve linear equations.

Solve each linear equation. Show how you know each answer is correct.

1. $x + 2x - \frac{1}{2}x = 15$ 2 points _____

2. $-4(2x - 0.5) = 10x - 1$ 3 points _____

3. $\frac{2}{5}(x + 10) + 3 = 2 + 2(\frac{1}{5}x + 5)$ 3 points _____

4. $15x - 3x = -\frac{3}{4}x$ 3 points _____

5. Charlie solves the following equation, but he makes a mistake.   3 points _____
   Circle the line where the error occurred and complete the solution correctly.

$$14x - 3 - 4x = 14 - \frac{1}{3}(6x - 3)$$
$$10x - 3 = 14 - \frac{1}{3}(6x - 3)$$
$$10x - 3 = 14 - 2x - 1$$
$$10x - 3 = 13 - 2x$$
$$12x = 16$$
$$x = 1\frac{1}{3}$$

- -

Reflection: Circle the sentence that best explains your understanding of solving linear equations. Explain your choice.

I can do this with some help.   I can do most of this.   I've got this!

Figure 3.2: Grade 8 common mid-unit assessment solving linear equations sample.

*Visit **go.SolutionTree.com/MathematicsatWork** for a free reproducible version of this figure.*

Common assessments given during a unit focused on one or two essential learning standards can provide you and your students with feedback related to the progression through the essential learning standards. This feedback, in turn, prepares students for the common end-of-unit assessment your team provided.

Assessments provided during the unit allow students the opportunity to re-engage in learning essential learning standards before the end-of-unit assessment. Ideally, you and your colleagues will work to erase inequities in learning as soon as possible as you ensure more students learn the essential learning standards in each mathematics unit.

TEAM RECOMMENDATION

Create Common Mid-Unit Assessments

- Create shorter common assessments to use during a unit that address one or two essential learning standards.

- Use the Assessment Instrument Quality Evaluation Rubric (figure 2.1, page 14) to determine the quality of the common mid-unit assessments.

How will you, your colleagues, and your students learn from the results of the common mid-unit assessment? What will you and your colleagues do before the unit ends if students have not yet learned the essential standards assessed?

The following are examples of end-of-unit common mathematics assessments that measure student learning in a unit, while also providing information you and your students can use formatively during the next unit.

Sample End-of-Unit Common Assessments

Consider the common end-of-unit assessment examples in figure 3.3 and figure 3.4 (page 47). The common end-of-unit assessment in figure 3.3 aligns to the grade 4 fractions unit essential standards in figure 2.3 (page 20).

TEACHER *Reflection*

You can use these questions to help your analysis as you *evaluate the quality* of each mathematics end-of-unit common assessment.

1. How well does each mathematics task align to the essential learning standards?

2. Is there a balance of lower- and higher-level-cognitive-demand tasks?

3. Is there enough variety between selected-response- and constructed-response task formats? Should you include technology, or do you not need it on this assessment?

4. How realistic and informative are the scoring agreements? Do you need to clarify the assigned point value or score for each mathematics task on the assessment?

5. Is the vocabulary throughout each assessment appropriate to the grade level or course?

6. Is there enough space for students to show their work?

7. How much time is necessary for students using this assessment? Is that time allotment acceptable, or do you need to modify accordingly?

_____

_____

_____

_____

_____

_____

_____

_____

_____

_____

_____

_____

_____

_____

Name: _____ Date: _____

Grade 4 Fractions Unit

I can explain why fractions are equivalent and create equivalent fractions.

1. Circle the two models that show equivalent fractions. Write the equivalent fractions.

2 points _____

$\bigcirc = \bigcirc$

2. Write two fractions that are equivalent to $\frac{6}{10}$.

2 points _____

$$\frac{6}{10} = \bigcirc = \bigcirc$$

3. Write two fractions that are equivalent to $\frac{10}{4}$.

2 points _____

$$\frac{10}{4} = \bigcirc = \bigcirc$$

4. Keisha says that $\frac{1}{2}$, $\frac{2}{4}$, and $\frac{3}{6}$ can all be worth the same amount.

2 points _____

   a. Do you agree with her? Circle yes or no: YES   NO

   b. Use sketches, numbers, and/or a combination of these to explain your thinking.

I can compare two fractions and explain my thinking.

5. Fill in the circle with <, >, or = to compare the two fractions.

1 point _____

$$\frac{1}{2} \bigcirc \frac{2}{3}$$

Figure 3.3: Team discussion tool—Sample grade 4 end-of-unit assessment.

continued →

6. Fill in the circle with <, >, or = to compare the two fractions.   1 point _____

$$\frac{3}{2} \bigcirc 1\frac{1}{2}$$

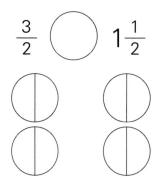

7. Fill in the circle with <, >, or = to compare the two fractions.   6 points _____

$$\frac{5}{5} \bigcirc 1 \qquad 1\frac{2}{3} \bigcirc 3\frac{1}{2} \qquad \frac{4}{5} \bigcirc \frac{3}{2}$$

$$\frac{7}{4} \bigcirc 1 \qquad \frac{5}{8} \bigcirc \frac{3}{4} \qquad 2\frac{2}{5} \bigcirc \frac{9}{4}$$

8. Mark says $\frac{1}{4}$ of his candy bar is smaller than $\frac{1}{5}$ of the same candy bar.   2 points _____
   He draws the following picture to explain his answer.

$\frac{1}{4}$			

$\frac{1}{5}$				

Mark's answer is wrong. Explain why Mark's answer is not correct, and then use pictures and words to explain the correct answer.

9. Explain why $\frac{4}{5}$ is greater than $\frac{2}{3}$. Be sure to include a picture in your explanation.   2 points _____

I can add and subtract fractions and show my thinking.

10. Complete the model and write the matching equation.   3 points _____

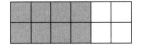 + 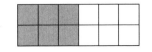 =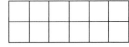
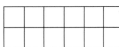

 + =

11. Complete the model and write the matching equation.   3 points _____

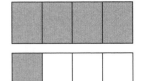 −  =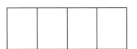

 − =

12. Show two different ways to add fractions that sum to $\frac{5}{8}$.   2 points _____

$$\frac{5}{8} =$$ $$\frac{5}{8} =$$

13. Find each sum or difference. Show how you know your answer is correct.   4 points _____

$$2\frac{2}{4} + 5\frac{3}{4} = \bigcirc$$ $$4\frac{1}{3} - 1\frac{2}{3} = \bigcirc$$

14. Charlie uses the following number line to show how to find $1\frac{2}{5} - \frac{3}{5}$. He made a mistake. Find his mistake and then find the correct answer to $1\frac{2}{5} - \frac{3}{5}$ and show how to find it on the number line.   3 points _____

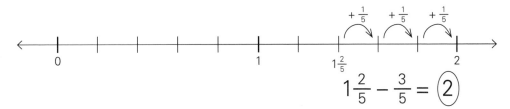

continued →

I can multiply a fraction by a whole number and explain my thinking.

15. Marta says the picture on the number line shows $\frac{1}{6} + \frac{1}{6} + \frac{1}{6} + \frac{1}{6} = \frac{4}{6}$. Jenny says Marta is correct and that the picture on the number line also shows a multiplication equation. Write the multiplication equation Marta sees in the picture.

2 points _____

16. Find each product.

3 points _____

$$7 \times \frac{1}{8} = \qquad\qquad 4 \times \frac{3}{4} = \qquad\qquad 5 \times \frac{1}{3} =$$

17. Complete the equation and show how you know your answer is correct.

2 points _____

$$2 \times \frac{5}{6} = \bigcirc \times \frac{1}{6}$$

I can solve word problems involving fractions.

18. Mark has five water bottles. Each water bottle holds one liter and is $\frac{2}{3}$ full. He pours them all into an empty water cooler. How many liters of water are in the water cooler now? Show how you know your answer is correct.

2 points _____

19. Jack is making pancakes. His mom said, "Add flour until the batter is hard to stir." First, he adds $1\frac{3}{4}$ cups of flour, and then he adds $\frac{2}{4}$ cup flour. Finally, the batter is hard to stir. How much flour did Jack add to the pancake batter? Show how you know your answer is correct.

2 points _____

20. After school, Carli goes for a run. She runs $1\frac{1}{2}$ miles, takes a quick break, and then finishes her run. Altogether, she runs $3\frac{1}{2}$ miles. How far did Carli run after her break? Show how you know your answer is correct.

2 points _____

Name: _____ Date: _____

Algebra 1/Integrated Mathematics I Linear Functions Unit

I can graph linear functions and interpret their key features.

1. Graph $f(x) = \frac{2}{3}x - 6$ and identify the slope, y-intercept, and x-intercept.   4 points _____

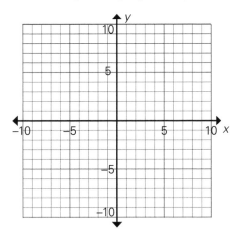

Slope: _____

y-intercept: _____

x-intercept: _____

2. Use the graph of the following function to determine $f(2)$.   1 point _____

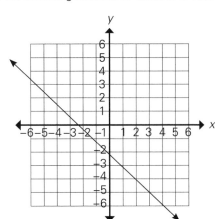

$f(2) =$ _____

3. Graph the line described by the linear equation $x - 6y = 18$, and identify the x- and y-intercepts. 3 points _____

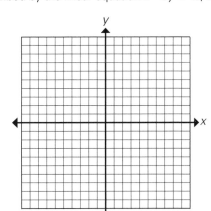

y-intercept: _____

x-intercept: _____

Figure 3.4: Team discussion tool—Sample high school algebra 1/integrated mathematics I end-of-unit assessment.

continued →

4. Determine the rate of change, *y*-intercept, and *x*-intercept of the following function.   3 points _____

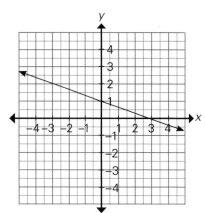

Rate of change: _____

x-intercept: _____

y-intercept: _____

5. Determine the *y*-intercept of each linear function.   2 points _____

Function A

x	0	5	10	15
y	15	5	−5	−15

y-intercept of function A: _____

Function B

$f(x) = 15x + 13$

y-intercept of function B: _____

I can calculate and explain the average rate of change (slope) of a linear function.

6. Determine the slope of each linear function.   2 points _____

Function A

x	0	2	6	10
y	6	7	9	11

Slope of function A: _____

Function B

$f(x) = -x + 2$

Slope of function B: _____

7. Determine the slope of the line shown in the following graph.   1 point _____

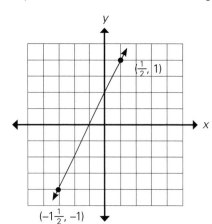

Slope: _____

8. The following two linear functions show the amount of money earned for each hour worked at two different stores. Use the slope of each function to explain at which store you would earn more money per hour.

3 points _____

	Groceries R Us				Buy It Here
Number of hours worked (x)	2	3	5	10	$9.5x - y = 0$
Amount of money earned (y)	18.50	27.75	46.25	92.50	

I can recognize situations with a constant rate of change and interpret the parameters of the function related to the context.

9. The function $f(x) = 40x$ describes how far from home Selena is as she drives from Dallas to Miami. Which graph best represents the function?

2 points _____

A.
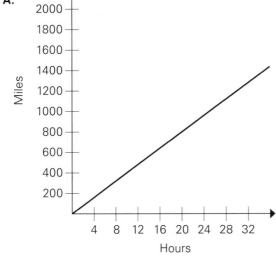

B.
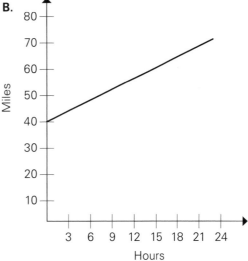

C.
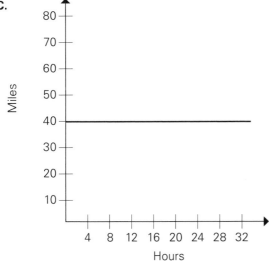

D.
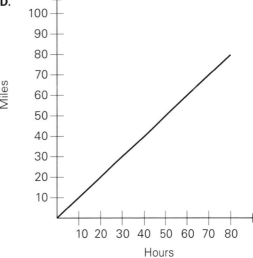

Using the information from your preceding answer, estimate how far from home Selena is in 8 hours.

continued →

10. The water level of a river is decreasing at a constant rate every day. The situation is represented by $w(d) = 34 - 0.54d$, where w is the water level in feet after d days.   4 points _____

   a. Determine the rate of change described in $w(d)$, and explain what it means in this context.

   b. Determine the y-intercept for $w(d)$, and describe what it means in this context.

11. When Briana has her picture taken, the photographer charges a $10 sitting fee and $6 for each sheet of pictures she purchases. This can be modeled using the function $f(x) = 6x + 10$, where x represents the number of sheets of pictures purchased. Briana has enough money to buy as many as 10 sheets of pictures.   3 points _____

   a. Determine the constant rate of change in this context.

   b. Determine the domain of this function given the constraints on Briana's amount of money.

   c. Determine the range of this function given the constraints on Briana's amount of money.

12. Martha drives from her house to school. She drives 5 miles, and it takes her 15 minutes.   2 points _____

   a. Determine Martha's average rate of speed.

   b. If you graphed Martha's distance traveled (miles) over time (minutes), will it be a linear function that matches Martha's average rate of speed? Explain why or why not.

I can construct linear functions given a graph, description, or two ordered pairs.

13. Don opens a savings account with $300. Each month, he will add $50 to the account.   2 points _____

   a. Write a function to model this situation where x is the number of months money has been added to the account.

   b. Use your function to determine how long it will take Don to save $1,050.

14. Brian has 64 flowers for a big party decoration. In addition, he is planning to buy some flower arrangements that have 18 flowers each, **if needed**. All the arrangements cost the same. Brian is not sure yet about the number of flower arrangements he wants to buy, but he has enough money to **buy up to 5 of them**.

2 points _____

a. Choose the function that best describes how many flowers Brian has for party decorations. Let x represent the number of flower arrangements Brian buys.

A. $f(x) = 18x + 64$

C. $f(x) = 64 + x$

B. $f(x) = 64x + 18$

D. $f(x) = 5x + 64$

b. Choose a reasonable **domain** for the function, keeping in mind that Brian may need to buy additional arrangements.

A. {0, 1, 2, 3, 4}

C. {1, 2, 3, 4, 5}

B. {0, 1, 2, 3, 4, 5}

D. {5}

15. The following graph shows the total cost of tickets (y) for a given number of rides (x) at a fair.

4 points _____

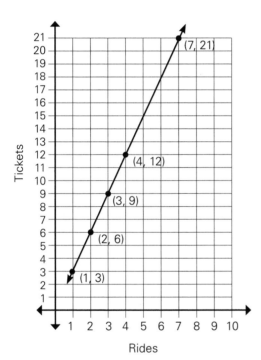

a. Write a function showing the cost for x rides.

b. Explain whether or not the model is accurate when drawn with a solid line.

c. Determine the largest number of rides James can go on if he has $35 to spend on ride tickets. Show how you know your answer is correct.

16. The cost of a prepaid cell phone plan from Cricket includes an activation fee plus a cost per minute. The total costs are represented in the following table.

5 points _____

Cricket

Minutes Used (t)	Total Cost $f(t)$
0	$7.50
1	$7.65
2	$7.80
3	$7.95
4	$8.10

T-Mobile also charges an activation fee plus a rate per minute. The total cost for T-Mobile's plan when t minutes are used is given by $g(t) = 10 + 0.10t$.

a. Write a function, $f(t)$, to show the total cost of the Cricket cell phone plan based on t minutes of usage.

b. Which company has the higher activation fee? Provide a mathematical justification for your answer.

c. Which company charges the most per minute of usage? Provide a mathematical justification for your answer.

continued →

17. Josh is planning a large anniversary party for his grandparents. He is deciding how many square tables to put together in a row to seat people. In his following drawings, each table is represented by a square, and each person is represented by a triangle.   5 points _____

  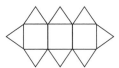 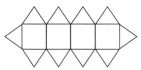

As the drawings show, Josh can seat 4 people around one table and 6 people around two tables put together.

   a. Complete the table to show how many people, p, can sit around t tables put together in a row.

Number of tables (t)	1	2	3	4	5
Number of people (p)					

   b. Write an equation that can determine how many people, p, can sit around a given number of tables, t.

   c. Use your equation to determine how many people can sit around 12 tables.

I can graph piecewise functions, including step and absolute value functions.

18. The graph represents the amount of water in Pam's rain barrel. Is each of the following a possible interpretation of the graph? Check *Yes* or *No* for each interpretation given below.   3 points _____

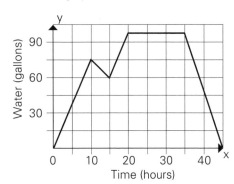

   a. The barrel starts off empty.
       ☐ Yes   ☐ No
   b. Pam uses water from the barrel to water her plants after 35 hours.
       ☐ Yes   ☐ No
   c. It rains for the first 10 hours.
       ☐ Yes   ☐ No

19. Given the absolute value function $g(x) = 3|x| - 5$:   4 points _____

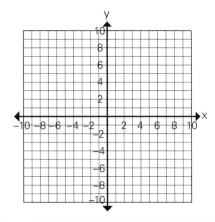

   a. Graph the function on the coordinate grid to the right.

   b. Identify the **vertex** _____.

20. The front of a camping tent can be modeled by the function $f(x) = -1.4\,|x - 3| + 4$ where the x- and y-axes are measured in feet. Using graphing technology, find the **maximum** height possible for the tent.

2 points _____

21. A bike rental company is having a summer special. The following equations show the prices charged for each hour of rental.

7 points _____

$$f(x) = \begin{cases} 10, & 0 < x < 1 \\ 20, & 1 \le x < 2 \\ 30, & 2 \le x < 3 \\ 40, & 3 \le x < 4 \\ 50, & 4 \le x \le 8 \end{cases}$$

a. Graph the cost of the bike rental for the amount of time the company rents the bike.

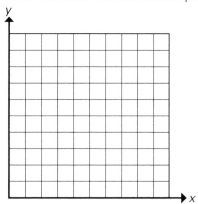

b. Based on the graph, how much will it cost to rent the bike for 3 hours?

c. If you have a budget of $20, determine the maximum amount of time you can rent the bike.

d. What is the maximum length of time possible for the bike rental?

The end-of-unit assessment in figure 3.4 (page 47) aligns to the high school essential learning standards for the linear functions unit in the online reproducible "Unit Samples for Essential Learning Standards in Mathematics." (Visit **go.SolutionTree.com /MathematicsatWork** for this free reproducible.)

Choose to examine either the fourth-grade assessment or the high school assessment. As a team, answer the questions, and then give the exam a high-quality score using the Assessment Instrument Quality Evaluation Rubric in figure 2.1 (page 14). All 4s in each criterion will result in a 32.

Did you rate the grade 4 fractions assessment as a perfect 32? Like any assessment, it can be improved. For example, this common assessment is most likely too comprehensive for the time allowed in class.

It seems reasonably balanced in terms of higher- and lower-level tasks, but at times may not leave enough space for student work. While the assessment clearly identifies the scoring agreements, there is not quite enough detail on the type of student work needed in explanations to be considered proficient.

If you serve on a fourth-grade team, you may also find some additional or replacement mathematical tasks you feel better meet the essential learning standards listed. That said, this is a strong assessment and certainly one that you can modify to be made even more productive for both teams and students. This is part of the professional work of your team every year.

Did you rate the algebra 1/integrated mathematics I linear functions assessment as a perfect 32? Like any assessment, collaborative teams can improve it. For example, this high school common assessment in figure 3.4 is most likely too comprehensive for the time allowed in class.

The common assessment seems reasonably balanced in terms of higher- and lower-level tasks, but again, like the fourth-grade assessment in figure 3.3 (page 43), may not leave enough space for student work. You might consider the clarity of directions for a single mathematics task or a group of task questions. Could you improve the directions for clarity?

If you serve on a high school team, you may also find some additional or replacement mathematical tasks you feel better meet the essential learning standards listed. That said, this is a strong assessment and certainly one that your team can modify to be even more productive

for it to use and students to learn from. This is part of the professional work of your team every year. (Visit **go.SolutionTree.com/MathematicsatWork** for additional unit assessment samples.)

You and your colleagues should take the time to examine all the current common mathematics unit assessment instruments you use with your students to determine strengths and weaknesses.

TEACHER *Reflection*

Compare the unit examples provided to your team's current unit mathematics assessments.

What are the strengths of your mathematics assessments, and in what areas do your assessments need improvement?

Do your colleagues agree with you? Why or why not?

_____

_____

_____

_____

_____

_____

_____

Once you establish your common assessments, examine sample assessments online or released items from state (or province) or national assessments to strengthen and improve your own common assessment instruments each school season. Writing good, high-quality common unit assessments should be an ongoing and improving aspect of your PLC culture. If you are a singleton teacher, consider using state or national released mathematics assessment tasks as a way to compare your students against a state or national student sample.

Equally important is reaching team agreement on how to score those tasks your team selects for your assessments. You should base the scores students receive for common unit assessments on the understanding you and your colleagues reach regarding the necessary work to receive either full or partial credit on each mathematics task.

Collaborative Scoring Agreements

Recall the Appropriate and Clear Scoring Rubrics section in chapter 2 (page 28). Consider these questions: Are the scoring rubrics used for every task clearly stated on your assessment? Does your scoring rubric total points or proficiency scale make sense based on the complexity of reasoning for each task? Most important, are the scoring points or proficiency scale assigned to each task appropriate and agreed upon by each team member?

In this section, you will explore another important step for your team's assessment work. Once you establish your common unit mathematics assessments, you and your colleagues must work together and find agreement on scoring those assessments. You must consider what demonstrations of learning you will require for each mathematics task on the assessments.

Your team's work will depend on whether you use scoring rubrics with a predetermined level of proficiency percentage expected, or proficiency scales as part of a standards-based grading system. We discuss both of these methods next.

Scoring Rubrics

In figures 3.3 and 3.4, the teacher team placed point values for each mathematics task or question on the assessment. How did the team decide the overall scoring procedures for the end-of-unit assessment? If you are a singleton teaching an advanced placement (AP) course, you may want to consult released exam questions to determine how to allocate points for scoring on your assessments.

Figures 3.5 and 3.6 (page 57) provide the *scoring* rubric the teacher team assigned for each mathematical task on the common end-of-unit assessments from figures 3.3 and 3.4. Choose the assessment closest to your grade level, and reflect on the scoring system this team decided to use for each mathematics task and each standard. Determine if you would accept it as is or change it based on your understanding of the essential learning standards.

Grade 4 Fractions End-of-Unit Assessment Scoring Rubric

1. I can explain why fractions are equivalent and create equivalent fractions.

Question	Total Points	Scoring Guidelines (Students must score 6 out of 8 points for proficiency.)
1	2	1 point per correct response (circle first two models: $\frac{3}{4} = \frac{6}{8}$ if they reference the shaded part and $\frac{1}{4} = \frac{2}{8}$ if they reference the unshaded part of the models)
2	2	1 point per correct equivalent fraction to $\frac{6}{10}$
3	2	1 point per equivalent fraction to $\frac{10}{4}$ (includes mixed numbers)
4	2	1 point for yes. However, student can earn one point total if he or she says no because the wholes might be different sizes. 1 point if student shows the three fractions are equivalent using the same-sized whole

2. I can compare two fractions and explain my thinking.

Question	Total Points	Scoring Guidelines (Students must score 9 out of 12 points for proficiency.)
5	1	1 point for <
6	1	1 point for =
7	6	1 point for each correct answer = < < > < >
8	2	1 point for noticing Mark drew different-sized wholes 1 point for explaining that $\frac{1}{5} < \frac{1}{4}$
9	2	1 point for pictures showing $\frac{4}{5}$ and $\frac{2}{3}$ 1 point for explaining how the pictures show $\frac{4}{5} > \frac{2}{3}$

Figure 3.5: Grade 4 fractions end-of-unit sample assessment scoring rubric.

continued →

3. I can add and subtract fractions and show my thinking.

Question	Total Points	Scoring Guidelines (Students must score 11 out of 15 points for proficiency.)
10	3	1 point for correct shading in final model ($\frac{14}{12}$ shaded) 1 point for correct addends ($\frac{8}{12} + \frac{6}{12}$) 1 point for correct sum ($\frac{14}{12}$, $\frac{7}{6}$, $1\frac{2}{12}$, or $1\frac{1}{6}$)
11	3	1 point for correct shading in the final model ($\frac{3}{4}$ shaded) 1 point for correct fractions on left side of equation ($\frac{5}{4} - \frac{2}{3}$) 1 point for correct difference ($\frac{3}{4}$)
12	2	1 point for correct sum to $\frac{5}{8}$ (such as, $\frac{5}{8} = \frac{1}{8} + \frac{1}{8} + \frac{1}{8} + \frac{1}{8} + \frac{1}{8}$) 1 point for correct sum to $\frac{5}{8}$ different from the first (for instance, $\frac{5}{8} = \frac{2}{8} + \frac{3}{8}$)
13	4	1 point per correct answer ($8\frac{1}{4}$, $2\frac{2}{3}$) for a possible total of 2 points 1 point per correct justification for a possible total of two points
14	3	1 point for correctly identifying the error (add instead of subtract) 1 point for correct answer ($\frac{4}{5}$) 1 point for showing how to find the answer on the number line

4. I can multiply a fraction by a whole number and explain my thinking.

Question	Total Points	Scoring Guidelines (Students must score 4 out of 7 points for proficiency.)
15	2	1 point for correct factors ($4 \times \frac{1}{6}$) 1 point for correct product ($\frac{4}{6}$)
16	3	1 point for each correct answer ($\frac{7}{8}$, 3, $\frac{5}{3}$ or $1\frac{2}{3}$)
17	2	1 point for correct answer in equation (10) 1 point for correct explanation

5. I can solve word problems involving fractions.

Question	Total Points	Scoring Guidelines (Students must score 4 out of 6 points for proficiency.)
18	2	1 point for correct answer ($\frac{10}{3}$ liters or $3\frac{1}{3}$ liters) 1 point for justification
19	2	1 point for correct answer ($2\frac{1}{4}$ cups) 1 point for justification
20	2	1 point for correct answer (2 miles) 1 point for justification

Visit **go.SolutionTree.com/MathematicsatWork** *for a free reproducible version of this figure.*

Algebra 1/Mathematics I Linear Functions End-of-Unit Assessment Scoring Rubric

1. I can graph linear functions and interpret their key features.

Question	Total Points	Scoring Guidelines (Students must score 10 out of 13 points for proficiency.)
1	4	1 point for correct graph of $y = \frac{2}{3}x - 6$] 1 point for correct slope: $\frac{2}{3}$ 1 point for correct y-intercept: $(0, -6)$ 1 point for correct x-intercept: $(9, 0)$
2	1	1 point for correct answer: $f(2) = -4$
3	3	1 point for correct graph of $y = \frac{1}{6}x + 3$ 1 point for correct y-intercept: $(0, 3)$ 1 point for correct x-intercept: $(-18, 0)$
4	3	1 point for correct rate of change: $-\frac{1}{3}$ 1 point for x-intercept: $(3, 0)$ 1 point for y-intercept: $(0, 1)$
5	2	1 point for correct y-intercept for function A: $(0, 15)$ 1 point for correct y-intercept for function B: $(0, 13)$

2. I can calculate and explain the average rate of change (slope) of a linear function.

Question	Total Points	Scoring Guidelines (Students must score 5 out of 6 points for proficiency.)
6	2	1 point for correct slope for function A: $\frac{1}{2}$ 1 point for correct slope for function B: -1
7	1	1 point for correct slope: 1
8	3	1 point for slope of Groceries R Us: $9.25 per hour 1 point for slope of Buy It Here: $9.50 per hour 1 point for recognizing you earn more per hour at Buy It Here

3. I can recognize situations with a constant rate of change and interpret the parameters of the function related to the context.

Question	Total Points	Scoring Guidelines (Students must score 8 out of 11 points for proficiency.)
9	2	1 point for correct answer A 1 point for using the graph or $f(x) = 40x$ for answer: About 320 miles
10	4	1 point for rate of change: -0.54 feet/day 1 point for explanation: The water is decreasing $\frac{1}{2}$ foot per day 1 point for y-intercept: $(0, 34)$ 1 point for explanation: Initially, the water level is at 34 feet
11	3	a. 1 point for rate of change: $6 per sheet of pictures b. 1 point for domain: {0, 1, 2, . . . 10} or $0 \le x \le 10$ where x is an integer (or whole number) c. 1 point for range: {10, 16, 22, . . ., 70} or $10 \le y \le 70$ where y is a multiple of 6 added to 10
12	2	a. 1 point for average rate of change: $\frac{1}{3}$ mile per minute b. 1 point for explaining that it is not a linear function since she did not travel at $\frac{1}{3}$ mile per minute the entire trip (such as, stopped along the way, started slower at first, slowed down at end, and so on)

Figure 3.6: Algebra 1/integrated mathematics I end-of-unit sample assessment scoring rubric.

continued →

4. I can construct linear functions given a graph, description, or two ordered pairs.

Question	Total Points	Scoring Guidelines (Students must score 13 out of 18 points for proficiency.)
13	2	a. 1 point for $f(x) = 50x + 300$ (or $y = 50x + 300$) b. 1 point for correct answer: 15 months
14	2	a. 1 point for correct answer A b. 1 point for correct answer B
15	4	a. 1 point for $f(x) = 3x$ (or $y = 3x$) b. 1 point for explaining that the domain only includes integers so the graph is not continuous (acceptable if mentioned that the line helps find total cost) c. 2 points possible: 　▸ 1 point for correct answer of 11 rides (with $2 left over) 　▸ 1 point for work or explanation to support the answer
16	5	a. 1 point for the function $f(t) = 7.50 + 0.15t$ b. 2 points possible: 　▸ 1 point for correct answer: T-Mobile has the higher activation fee 　▸ 1 point for comparing T-Mobile activation fee of $10 to Cricket of $7.50 c. 2 points possible: 　▸ 1 point for correct answer: Cricket has the higher rate per minute 　▸ 1 point for comparing T-Mobile rate of $0.10 per minute to Cricket's $0.15 per minute
17	5	a. 2 points for correctly filling in the table with the values 4, 6, 8, 10, 12 (1 point if at least three are correct) b. 2 points for correct equation: $p = 2 + 2t$ c. 1 point for correct answer: 26 people

5. I can graph piecewise functions, including step and absolute value functions.

Question	Total Points	Scoring Guidelines (Students must score 12 out of 16 points for proficiency.)
18	3	a. 1 point for yes b. 1 point for yes c. 1 point for yes
19	4	a. 3 points for correct graph: 　▸ 1 point for vertex at (0, −5) 　▸ 1 point for left side of graph 　▸ 1 point for right side of graph b. 1 point for vertex (0, −5)
20	2	2 points for sketch and correct answer: 4 feet
21	7	a. 4 points for correct graph, specifically: 　▸ 1 point for labels 　▸ 1 point for closed points on the left of each "piece" 　▸ 1 point for each open point on the right of each piece 　▸ 1 point for connecting each piece in horizontal segments b. 1 point for correct answer $40 c. 1 point for correct answer: up to 2 hours (1 hour 59 minutes) d. 1 point for correct answer: 8 hours

*Visit **go.SolutionTree.com/MathematicsatWork** for a free reproducible version of this figure.*

You may have noticed in the two samples in figures 3.5 (page 55) and 3.6, that each teacher team established proficiency for each essential learning standard on the assessment by setting total-points-correct expectations for each standard.

This type of proficiency setting requires acceptance that if a student meets the total number of points correct (for example, 9 of 12), then he or she has demonstrated the expected learning for that standard. Another approach is to use proficiency scales.

Proficiency Scales

Some collaborative teams use a *proficiency scale* to score student work for essential learning standards. Teachers use the scale to determine if the student has one of the following for each essential learning standard.

- Minimal understanding

- Partial understanding

- Proficient understanding

- Above grade-level understanding

A proficiency scale can be a good way to measure student progress on a standard over time and during several units of study. This is done most often in the elementary grades but you can use it in any K–12 grade if you identify the essential learning standards to monitor. You can record progress from one grading period to the next.

Figure 3.7 shows an example of a proficiency scale for the first-grade common mid-unit assessment in figure 3.1. Notice that it is not a scale for each question on the assessment, but rather an overall proficiency rating related to the essential learning standard. On an end-of-unit assessment, there would be one proficiency scale per essential learning standard. Also notice that students cannot earn a level 4 on this assessment since all task questions are at grade level.

If you and your colleagues use proficiency scales similar to the one in figure 3.7, it is important to identify the quality of student work expected at each proficiency level as it relates to the standard rather than inserting points into the scale. It is also important to determine whether the end-of-unit common assessment is the best place to assess students at level 4 or whether students can demonstrate learning at level 4 through formative assessment experiences during the unit or with work on a separate level 4 common assessment. On page 60, author Sarah Schuhl describes her recent experiences

Directions: Compare the proficiency scale for the place value check with the assessment in figure 3.1 (page 40). How would you and your teammates interpret each score? What would a student have to show to earn a 1, 2, or 3?

Level 1 Minimal Understanding	Level 2 Partial Understanding	Level 3 Proficient Understanding	Level 4 Above Grade-Level Understanding
Student can sometimes identify the largest number when comparing two numbers.	Student can identify the greater than or less than number and use the correct symbol for comparison, but cannot explain the comparison using place value. Or Student can identify the greater than or less than number and can often use the correct symbol for comparison with a minimal ability to explain the comparison using only ones (not tens as a group).	Student can identify the greater-than or less-than number and use the correct symbol for comparison (with a possible simple mistake for an error) and can explain the comparison using place-value language.	Not applicable for this assessment; needs additional data. Ask student to compare three-digit numbers and explain why one number is larger than another using place value language or have students identify a number that is greater than a given number and less than a second number, and write the entire inequality using two inequality symbols.

Figure 3.7: Proficiency scale for grade 1 common mid-unit assessment sample in figure 3.1.

Personal Story 🟣 **S A R A H   S C H U H L**

In my work with teams, I have learned that proficiency scales are not always easy to create, but they provide rich opportunities for teachers to talk about what students have to know and be able to do to be proficient with the mathematics standards.

At a recent high school collaborative team meeting with a team transitioning to proficiency-scaled assessments, the teachers wanted my opinion on their scale. The scale did not reference a total number of points needed for the essential learning standard portion of the assessment. Interestingly, the scale also did not reference student learning levels related to the standard. Instead, the proficiency scale identified which questions a student had to answer correctly to earn a score of 1, 2, 3, or 4. For example, students earned a 1 if they answered questions 1 and 2 correctly, a 2 if they also answered question 3 correctly, and a 3 if they also answered question 4 correctly. Question 5 was beyond the intent of the standard, so if students answered all five questions correctly, they earned a 4 on the proficiency scale. On such a scale, what happens if a student only answers question 3 correctly? Or question 5? What happens if a student misses the first question and the rest are perfect?

As we wrestled with the options, it became clear the team was going to need to create a scale showing the type of learning evidence a student produces for each level 1–4. We looked at a rubric from ThemeSpark (www.themespark.net) and revised it as needed, noting that a proficient student might still make some simple mistakes and need additional practice, but is not in need of re-teaching.

The team also decided not to assess level 4 on the end-of-unit common assessment but rather as a separate test. This would allow them to ask more questions at grade level and make sure students did not spend too much time on questions beyond the intent of the standards at the expense of having time to demonstrate learning on other essential standards. Instead, the team created five questions, some at a low level and others at a high level (but not beyond the intent of the essential mathematics standards). Team members did this only *after* they made their proficiency scale, so they could discuss the type of work expected for a student to earn a score of 1, 2, or 3 using the preponderance of evidence produced by the student on that portion of the exam.

developing proficiency scales. Sarah's experience with this high school team is similar to PLC team discussions for elementary and middle teachers.

Whether you use total points or proficiency scale scores, you and your mathematics teams should define scoring agreements and calibrate the scoring agreements for all your common assessments.

Finally, to support your collaborative scoring work, choose one of your own end-of-unit common assessments, preferably one you will be using in the next few units. Use the tool in figure 3.8 to determine how you will score the items on the assessment and how you will know whether or not students are proficient with each essential standard.

There is no right or wrong way to determine the points for scoring exams. What is important is that you and your colleagues can use the same scoring scale and base the proficiency scale score or points for each task on a decided standard (such as the complexity of reasoning required by the assessment task or a proficiency scale based on lower- or higher-level cognitive demand).

Reaching these agreements will create a greater likelihood of erasing any student inequities in your individual scoring and grading practices once students take your common assessments. There is also an increased fidelity to your answer to PLC critical question 2 (DuFour et al., 2016): How will we know if students learn it?

Assessment Instrument Alignment and Scoring Rubric

Directions: Within your collaborative team, answer each of the following eight questions in relation to your team common assessment.

1. Which essential learning standard does each task address, and how do you know that the task aligns to the essential learning standard?

2. What work do you expect students to demonstrate in order to successfully respond to and receive full credit for each task on the assessment?

3. How will you assign partial credit for each task?

4. Which mathematical practices or process standards does the task develop?

5. What scoring value or point value would you assign to each task or group of tasks?

6. If you score with points, how many total points should you assign to this end-of-unit assessment? If you score using a proficiency scale, how will you score students for each essential learning standard?

7. Are there any questions on the test you would want to ask differently? If so, how would that affect the point value or proficiency scale score you assign?

8. How many points correct would a student need for each essential learning standard to qualify as proficient for that standard (the proficiency target)? Or, what must a student demonstrate to earn a 3 for each essential learning standard if you use a proficiency scale score?

Figure 3.8: Team discussion tool—Assessment instrument alignment and scoring rubric.

*Visit **go.SolutionTree.com/MathematicsatWork** for a free reproducible version of this figure.*

That answer lies partially in the hard work you do as a team to reach scoring consensus with your colleagues.

As your collaborative team begins the assessment work within a PLC culture, you will pursue building common unit assessments. Often removed from the team discussion for common assessments is the team deliberation and agreement on the appropriate scoring of student work for each mathematics problem or task on the assessment. The criteria for scoring, as described in this chapter, will help you, your students, and your students' families to gain confidence that the score is accurate and provides the necessary feedback for the formative purposes described in part 2.

TEAM RECOMMENDATION

Create Common End-of-Unit Assessments

- Create common assessments to use at the end of each unit that address the three- to six-unit essential learning standards. Ideally, create this before the unit is taught.

- Using the eight criteria in figure 2.1 (page 14), Assessment Instrument Quality Evaluation Rubric, determine the quality of the common end-of-unit assessments.

Part 1 of this book discussed the assessment development work necessary to help you and your collaborative mathematics team improve your response to the first two PLC critical questions (DuFour et al., 2016).

1. What do we want all students to know and be able to do?

2. How will we know if students learn it?

You and your colleagues can address these first two PLC critical questions through your efforts to work on team action 1: Develop high-quality common assessments for the agreed-on essential learning standards. To achieve team action 1, you and your colleagues should work together to reach agreement on the following five factors for each grade level or course for mathematics.

1. The essential learning standards for mathematics

2. The daily learning targets that result from unwrapping the standards

3. The rigor level of all unit assessment tasks based on the conceptual understanding, application, and procedural fluency students need to demonstrate proficiency for each essential learning standard

4. The development of high-quality common unit mathematics assessments, both during and at the end of each unit

5. The development of common scoring agreements for each mathematical task on the common assessments

Ideally, you complete these teacher team assessment artifacts before the unit begins to bring clarity and focus to your instructional response to student learning throughout the unit. Metaphorically, it is as if your team selects the beach for learning and then designs a high-quality surfboard for that beach.

Professionally, why go through all this work? Because as W. James Popham (2011) reminds us, the victory with your students is to use those great assessment instruments, work with your colleagues to ensure you are scoring them accurately, and then use them as part of a *formative process* with your students—an intervention and reflective process that demands meaningful feedback and action. You will explore this process in part 2.

Recall the original question in chapter 1 (page 7): What is the purpose of common unit assessments? Reaching team agreement for the essential learning standards on each unit assessment will support student action on your feedback. Reaching team agreement on appropriate scoring rubrics for the assessment instrument reveals your equitable expectations for accurate evidence of student learning.

Thus, the value of any collaborative team–driven common unit mathematics assessment depends on the extent to which the assessment instrument:

- Reflects the essential learning standards and clearly indicates those standards on the assessment instrument in student-friendly *I can* language

- Provides valid evidence of student learning for each essential standard

- Results in a positive impact on student motivation and continued learning

- Supports a student formative process for learning after the assessment is returned

Now that you and your colleagues have created high-quality common assessments with common scoring agreements with all the teachers on your team, it is time to think about how you will use the common assessments for continued student learning. How will each assessment be a meaningful learning tool for students? How will each during- and end-of-unit mathematics assessment become a valuable part of the formative learning process for you and your students?

In part 2, you will examine how to use common assessments with students for continued learning of the essential learning standards. Each common mathematics assessment is only as valuable as the timely feedback it provides to you and your students and the opportunities it uncovers for targeted interventions and extensions.

Team Action 2:
Use Common Assessments for
Formative Student Learning and Intervention

In a PLC culture, you ensure that your collaborative team implements a required
response to intervention for mathematics.

—*Timothy D. Kanold and Matthew R. Larson*

What happens in the classrooms across your collaborative team when you return a common mathematics assessment to your students? What do you expect students to do with the feedback you provide? The answer to these questions is the purpose of part 2 and results in the most important benefit of working together as a team to design the high-quality assessments discussed in part 1. It is only in the nature of the expected student response to your feedback during and at the end of the unit that your unit assessments become *formative* for students. And it is through the formative feedback process that your intervention work as a team resides.

Remember, as part of a PLC culture, you and your colleagues are expected to embrace a high-quality intervention response to the third and fourth PLC critical questions (DuFour et al., 2016).

3. How will we respond when some students do not learn?

4. How will we extend the learning for students who are already proficient?

In short, how do you, your team, and your students use assessment results and evidence from student work to intervene and continue learning the essential standards for each mathematics unit? This formative assessment and intervention process is the focus of team action 2: Use common assessments for formative student learning and intervention.

In part 2 of this book, you will encounter several team discussion tools to support your team's formative assessment and intervention work.

*   Formative Use of Common Assessments (figure 4.1, page 68)

*   Common Unit Assessment Formative Process Evaluation (figure 4.2, page 70)

*   Equitable Scoring and Calibration (figure 4.3, page 73)

*   Intervention Practices (figure 5.1, page 87)

*   Mathematics Intervention Planning Tool (figure 5.3, page 91)

*   Collaborative Team Feedback With Student Action (figure 6.4, page 105)

In addition, you will explore what you and your colleagues should do once you administer a mathematics assessment, score the assessment, and pass it back to students.

In other words, *now what?*

Now that the assessment is written:

- How will it frame instructional support and student involvement in their continued learning?

- How will the results shape a team instructional response, and how will students be required to learn the content they have not learned *yet*?

- What will students do and what will you do to make sure the common assessment instrument is a tool that enhances student learning?

The professional work of your team begins with a close examination of the nature of highly effective feedback to students.

CHAPTER 4

Quality Mathematics Assessment Feedback Processes

Assessment is a process that should help students become better judges of their own work, assist them in recognizing high-quality work when they produce it, and support them in using evidence to advance their own learning.

—*NCTM, Principles to Actions*

There are two important elements to consider when using your mathematics assessment results effectively. First, students must receive *meaningful impact* feedback on their work in order to improve their mathematical understanding. For the purposes of this book, meaningful assessment feedback refers to the FAST feedback elements described in this chapter.

Second, students must use tools or protocols requiring them to take action on during-the-unit and end-of-unit mathematics assessment feedback in formative ways. To repeat: they *take action* on your feedback. Dylan Wiliam (2016) states it this way, "If our feedback doesn't change the students in some way, it has probably been a waste of time" (p. 14).

Effective and meaningful feedback must have four specific characteristics to be meaningful to students and allow for them to grow in their learning. Remember, the purpose of this book and of your professional work as a mathematics teacher is that *every student can learn mathematics*.

FAST Feedback

The Mathematics in a PLC at Work team uses the acronym *FAST* (*fair, accurate, specific,* and *timely*) to describe the essential characteristics of meaningful formative feedback explained by Reeves (2011, 2016) and Hattie (2009, 2012). FAST feedback works well when using evidence of learning from common assessments during or at the end of the unit. FAST feedback includes the following characteristics.

- **Fair:** Effective feedback on the mathematics assessment rests solely on the quality of the student's demonstrated work and not on other student characteristics.

- **Accurate:** Effective feedback on the mathematics assessment acknowledges what students are currently doing well and correctly identifies errors they are making. According to Stephen Chappuis and Rick Stiggins (2002), "Effective feedback describes why an answer is right or wrong in specific terms that students understand" (p. 42).

- **Specific:** Your mathematics assessment notes and feedback "should be about the particular qualities of [the student's] work, with advice on what he or she can do to improve, and should avoid comparison with other pupils" (Black & Wiliam, 2001, p. 6). Try to find the right balance between being specific enough so the student can quickly identify the error in his or her logic or reasoning but not so specific that you end up correcting the work for him or her. Does your assessment feedback help students correct their thinking as needed?

- **Timely:** You provide effective feedback on the end-of-unit mathematics assessment just in time for students to take formative learning action on the results before too much of the next unit takes place. As a general rule, you should pass back during-the-unit and end-of-unit assessments with expected results for proficiency to students within forty-eight hours of the assessment.

As the Wiliam (2016) quote mentioned, unless feedback *grows the learning* of students and changes or

TEACHER *Reflection*

What do you currently expect your students to do when you return an assessment to them either during or after a unit of study? Describe what currently happens in class with your students when you pass a graded mathematics assessment back to them.

_____

_____

_____

_____

_____

_____

_____

_____

validates their understanding, it has been a waste of time. What students *do* with your FAST feedback is the main idea of part 2 of this book.

How you create additional student learning opportunities from the common assessment feedback students receive is what allows you to use your assessment results as part of a *formative* learning process. The assessment becomes much more than an *event* for students. It becomes a valuable tool in the learning *process*.

Use the discussion tool in figure 4.1 to determine your team's current reality related to using high-quality FAST feedback with your common assessments.

Feedback is only effective if students act on it and take ownership of their learning. Effective feedback allows students to reflect on and understand their progression in demonstrating learning of the essential mathematics standards for each unit and identify actions they need to take to advance their learning.

Directions: Use the following prompts to guide team discussion of your current use of FAST feedback for all common during-the-unit and end-of-unit mathematics assessments.

Feedback to students:

1. How quickly do students receive their assessment results from each teacher on the team?

2. How do you provide corrective feedback to student errors on the common assessment questions?

Student actions:

3. What do you expect students to do with the assessment when you return the assessment?

4. How do you require students to reflect on their learning based on evidence from assessment results?

Teacher intervention actions:

5. How does your team help students re-engage in learning content?

6. How does your team analyze trends in student work (for example, common misconceptions) when discussing common assessment results? What do you do with that information?

7. What does your team collectively do to help students re-engage in learning content?

Figure 4.1: Team discussion tool—Formative use of common assessments.

*Visit **go.SolutionTree.com/MathematicsatWork** for a free reproducible version of this figure.*

Evaluation of Mathematics Assessment Feedback Processes

Throughout the feedback process, you and your colleagues should ask students to reflect on the evidence of their learning, refine their understanding of mathematical concepts and applications, and act on that knowledge to demonstrate proficiency. Along the way, you and your students use the assessment instruments as learning tools within a formative feedback loop.

The following personal story from author Jessica Kanold-McIntyre highlights the benefits of using a team-developed student feedback process with the first-grade students in her district.

You can use figure 4.2 (page 70) to identify areas to improve in your teams' formative assessment process. The action students take on your FAST feedback and your coordinated team efforts at designing equitable quality interventions impact every student's mathematics learning. We explain the first two criteria from figure 4.2 in chapter 1. You should evaluate and discuss the progress of your team on these two criteria to ensure they exist at a level 3 or 4, per figure 4.2. Agreeing on the essential learning standards and the common assessments of those standards is an essential linchpin to your teams' professional formative assessment work related to equitable learning expectations for every student.

Take a moment to use figure 4.2 to evaluate your current team effort to use your common assessments as part of a meaningful formative feedback process with your students. Give your team a score of 1 to 4 for each of the six criteria.

After evaluating your common unit assessment formative process, what did your team discover? How did you score?

Formative feedback for students is most effective when carefully planned. Intentional planning includes determining the essential learning standards and creating common unit assessments that reveal student thinking and learning. From the data revealed through student work on the assessments, your team can plan for students to reflect and set goals for continued learning. You can also plan for students to re-engage in learning through the intervention opportunities your team provides.

Personal Story **JESSICA KANOLD-McINTYRE**

There is a great power in the calibration process! I worked with a first-grade team who had been giving common assessments for a while. However, the teachers had just started talking about how they give feedback on assessments and how students reflect on their assessments by learning target. The team had developed a feedback form for students to use when they passed back the common mathematics assessments. The teachers were so excited to use the feedback form with their students.

At our meeting, when asked how the feedback form was working, the five teachers comprising the team realized that they all used the form in a different way. Until that moment when they heard each other share their expectations for student responses to errors on each standard, they had not realized they each used the form differently and at different times. The teachers realized they had all assumed everyone interpreted the expectations similarly and were surprised by the unintentional inequities they were causing their students. The conversation became an amazing collaboration in which they worked together to come to consensus on each part of FAST feedback and created a team Tier 2 intervention response to the assessment they could all share. Their willingness to discuss how to use the form to best support student learning required each of them to also approach the conversation in a professional and respectful manner.

Team Common Formative Assessment Process Criteria	Description of Level 1	Requirements of the Indicator Are Not Present	Limited Requirements of This Indicator Are Present	Substantially Meets the Requirements of the Indicator	Fully Achieves the Requirements of the Indicator	Description of Level 4
1. Agreed-on essential learning standards for the unit	Essential learning standards are unclear or differ among teachers on a team.	1	2	3	4	Essential learning standards are clear and commonly worded for students and shared with students at the start and throughout the unit.
2. Common unit assessments	Teachers on the team give their own mid-unit assessments and end-of-unit assessments.	1	2	3	4	Teachers design and administer common unit assessments, analyzing and responding together to the student results by student and essential learning standard.
3. Calibration of scoring agreements and student feedback	Teachers on the team score the assessments (common or not) individually, and each gives his or her own form of feedback to students.	1	2	3	4	Teachers regularly double score assessments to verify accurate scoring of student work (calibration) and determine the best way to provide feedback to students.
4. Student self-assessment and action after the end-of-unit assessment	Teachers do not ensure each student reflects on learning after the end-of-unit assessment to identify what the student has learned or not learned yet and to make a plan for future learning.	1	2	3	4	Team creates a system for students to self-assess by essential learning standard what the student has learned or not learned yet and creates an action plan to re-engage students in learning standards within team-created structures.
5. Student self-assessment and action from common mid-unit assessments	Teachers do not ensure each student reflects on learning after common mid-unit assessments during a unit to identify what the student has learned or not learned yet and to make a plan for future learning while still in the unit.	1	2	3	4	Team creates a formative system for students to self-assess by essential learning standard from common mid-unit assessments to identify what the student has learned or not learned yet, and creates an action plan to re-engage students in learning within team-created structures during the unit.
6. Team response to student learning using Tier 2 intervention criteria	Teachers on the team each determine how they provide students opportunities for intervention. Teachers on the team design interventions independently of one another.	1	2	3	4	Team develops a collective, just-in-time response to student learning by student and by essential learning standard, creating structures and plans for students to re-engage in and demonstrate learning as a result of teacher team actions.

Figure 4.2: Team discussion tool—Common unit assessment formative process evaluation.

How did you score your grade-level or course-based mathematics team on the six criteria from figure 4.2? How did your team do? Describe the most urgent assessment issue you need to address with your colleagues related to your formative use of common assessments.

_____

_____

_____

_____

_____

_____

_____

Quality of Mathematics Assessment Feedback

Student action from the results of your common assessments hinges on the nature of your consistent feedback from teacher to teacher on your team. This will, over time, erase any potential inequities for student learning opportunities.

The ability and willingness of your team to calibrate scoring on common assessments allow for a strong and shared intervention program and student re-engagement in learning and goal setting. First, the team scores and calibrates feedback for common end-of-unit assessments, and then students act on your subsequent feedback.

Calibration of Scoring Agreements and Student Feedback

Suppose your team gives a common assessment on Tuesday and then plans to review the results with students on Thursday. Unfortunately, life has been hectic and grading your assessments in time seems like an enormous challenge.

If a colleague knows of your situation and offers to grade your common unit assessments for you, how would you respond? Would you feel relieved and grateful for the help and support, or would you be panicked at the thought that he or she might not score them the same as you?

In chapter 1 (page 7), you identified common scoring agreements when designing the tasks for each unit assessment. Once you give the common assessment to students, there is an urgent moment when your team meets to calibrate your scoring and receive feedback from each other on your scoring accuracy for the common unit assessments. This is urgent because you need to return the feedback in a timely manner (recall the T in FAST feedback means timely), and you will need to reduce the potential for wide variability in how you and your team members score similar assessment tasks and questions.

It is in this moment (after students take an assessment) that you can agree on the scoring and accuracy of the feedback you give to students. If you do not agree, grading and scoring inequity will occur within the grade level or course. Scoring student work together provides insight into other team members' mathematical thinking and allows you to consider multiple ways you and your students represent their thinking.

Do you currently double score your unit assessments with a colleague? In other words, do you grade each other's student work? If so, where and when do you get it done?

_____

_____

_____

_____

_____

_____

Fidelity to the formative feedback process occurs when your team meets to score student work fairly and consistently. To strengthen this practice of consistently scoring student work, your team should practice common scoring of the same tasks at the end of each unit throughout the year. As the story from author Sarah Schuhl (page 72) indicates, the results of your initial calibration work together may be surprising.

Personal Story 66 **SARAH SCHUHL**

When I first asked our algebra team to collectively score a common unit assessment on linear equations, team members indicated it would be a waste of time. They had written the test together as well as how many points each item was worth on the assessment. However, they agreed to try.

We met together and scored a student assessment, only to learn the student would have earned an overall score of D– to B+, depending on which of the five teachers on the collaborative team graded the test. We were shocked!

If we had handed back assessments for reflection and goal setting without calibrating our scoring, some students would be told they needed intervention and others told they met the essential learning standards, depending on which teacher might have scored the assessment.

This disparity meant that using the same assessment to the same standards is not enough to ensure *equitable outcomes* for students across each teacher on our team with respect to consistently grading student work on our common assessments.

To view common assessments through more of an equity lens, students should receive similar scores from each team member (within one or two points for the entire end-of-unit exam or the same proficiency scale score for each essential learning standard). This also translates into similar scores for each essential learning standard on the assessment.

Some more common methods for ensuring scoring agreements include the following.

- **Double scoring:** This means one teacher scores exams and then a second teacher also scores the exams, placing both grades on the assessment. When there are variances, a third scorer or grader (usually a teacher on the team or a teacher who knows the mathematics content standards for the unit assessment) may be brought in to also score the exam, or the teachers work to clarify and agree about how to commonly score the essential learning standard or task in question.

- **Calibration:** This is when the teachers agree on the points or rubric to use to score papers and use anchor papers to gain clarity about student work and its interpretation related to proficiency.

- **Inter-rater reliability:** This occurs when two or more teachers score the same paper and examine their results during blind double scoring. Over

time, and as each teacher engages in discussions about how he or she arrived at the final assessment score, he or she resolves any significant differences and begins to assign student scores with feedback that is reasonably close. Thus, there becomes a greater score reliability across teachers scoring (or rating) the common assessment.

When you calibrate scoring as a team, you also calibrate your own understanding of the standards and proficiency expectations, which translates into accurate interpretation of results for teachers and students alike.

If you give a common assessment in a one-on-one interview setting and record student responses (such as in kindergarten), your team may need to video a few of your assessment interviews for the purpose of calibrating your scoring. You can use the videos in a team meeting to calibrate your interpretations of student learning as evidenced in real time. How closely and accurately do you and your colleagues score student evidence of learning?

You can use the team activity in figure 4.3 to facilitate a collaborative scoring conversation around either a recent end-of-unit common assessment your team gave or an upcoming assessment for the end of a unit. In a few instances, you may need to bring in videos from common assessments rather than student papers.

Work with your colleagues to determine the best methods for providing feedback to students that promotes the team's desire for continuous student learning. This starts with calibrated scoring agreements on both shorter, mid-unit common assessments and end-of-unit common assessments. Students know their score will not change if another teacher on the team grades the assessment.

The information the team gleans from each assessment is more useful when it is done by essential learning standard so all stakeholders can determine what students have learned and not learned *yet*, providing hope for improvement in students' and teachers' minds alike.

Once team members work through the calibration activity in figure 4.3, it is important to continue addressing assessment scoring calibration together with future unit common assessments. You might also consider rotating your scoring partner as you seek a more equitable and calibrated student feedback response on your common assessments.

If you are a singleton, you may need to use the calibration activity in figure 4.3 with other colleagues in your department to ensure calibrated grading vertically, so students know feedback is consistent from one year to the next. You might also calibrate by grading released student work from a state (or province) or national assessment and comparing your scoring to that of the released examples. Finally, you might engage with a virtual team, in which case you can engage in calibrating your scoring and feedback electronically. It is critical that students experience equity in their scoring feedback for optimal continued learning.

Beyond the scoring of the assessments, how does your team ensure the assessment feedback is FAST—fair, accurate, specific, and timely? How is your team working to ensure students can learn from the feedback so learning does not stop with a single score on an assessment?

John Hattie, Douglas Fisher, and Nancy Frey (2017) remind us that "what we say to students, as well as how we say it, contributes to their identity and sense of agency, as well as to their success" (p. 203). Further, they find there is a significant impact on student

---

Directions: Choose one other team member and plan to meet for thirty minutes at the end of the day. Each brings six to eight student papers or assessments to grade and score together. Use the following five steps to calibrate your scoring agreement and student feedback.

1. Bring student assessments that you believe will show strong evidence of student learning (proficient or advanced), partial evidence of student learning, and minimal evidence of student learning. In other words, those assessments that might rank high, medium, and low.

2. Blind double score the assessments using your agreed-on scoring rubrics, not letting your scoring partner know the score you give the student work or the feedback you would provide.

3. Compare scores and determine if there are any score variances for any standards or areas of the assessments.

4. Examine the mathematical tasks on the assessment that caused the greatest variance, and resolve those issues.

5. Once you calibrate your scoring expectations for student work with a teacher partner, score the rest of the assessments on your own using the feedback you receive from your colleague and the agreements you generate together.

Figure 4.3: Team discussion tool—Equitable scoring and calibration.

*Visit **go.SolutionTree.com/MathematicsatWork** for a free reproducible version of this figure.*

learning (an effect size of 0.75) through frequent and high-quality formative feedback on assessments (Hattie et al., 2017). The end-of-unit assessment is part of that process as students continue to learn throughout the next unit of mathematics.

TEACHER *Reflection*

Remember, the use of accurate and specific feedback only helps students if it is also *timely*. In general, how quickly does a student receive feedback from you on a mid-unit or end-of-unit assessment performance? What are the greatest barriers you face to providing timely feedback, and how can your team improve in this important part of the common mathematics assessment process?

_____

_____

_____

_____

_____

_____

_____

_____

_____

_____

_____

_____

_____

As an example, imagine a student only provides one solution pathway to solve a problem, and you provide feedback hints on the assessment task that could guide the student to consider another approach, such as, "What related facts do you know that could help you?" or "How could you use a graphing strategy?" If you provide feedback to the student well after you give the assessment, your feedback will have minimal impact. Students too often forget what they were thinking when they took the exam and tend to start over instead of working to revise their thinking and reasoning for greater learning impact.

TEAM RECOMMENDATION

Calibrate Scoring of Common Unit Assessments

- Double score common assessments to calibrate feedback to students.

- Discuss how to give FAST feedback to students within a quick time frame.

- Determine how you and your colleagues will calibrate scoring and explore effective ways to give meaningful feedback to students that is consistent from teacher to teacher.

Once you calibrate scoring of common assessments and discuss the nature of quality feedback, you should think about team structures that *require* students to take action on that feedback for continued learning, which we discuss in the following section.

Student Action After the End-of-Unit Assessment

When you return student work with your feedback on the end-of-unit assessment, your students need to self-assess if they are meeting the proficiency targets set for each essential learning standard for that unit. Your FAST feedback allows students, with your support, to engage in activities that lead to learning the essential learning standards for that unit, despite moving on and into the next mathematics unit for learning.

How will your team provide quality feedback to students and then *require* them to take action on your feedback as part of their continuous learning?

Students can articulate what they have learned and what they have not learned as early as prekindergarten or kindergarten and throughout high school. Students can reference their learning using learning standards if you share those standards at the beginning of a unit, throughout the unit within instruction as part of homework, and during reflection and planning after an end-of-unit common assessment.

Personal Story BILL BARNES

Elizabeth Kunstman, a mathematics team leader at Edison Middle School in Green Bay, Wisconsin, and her team were finding the commitment to high-quality student reflection and goal setting to be incredibly time-consuming and cumbersome. Providing assessment feedback to students, monitoring student analysis of strong and weak performing standards on the assessment, and ensuring student response to re-engage in learning required a massive paper trail for teachers.

Elizabeth's team members figured out an efficient way to use technology to help them with this aspect of their formative learning work. Each team member asked his or her students to utilize the Google Sheets tool to record returned assessment scores by standard. This process enabled students to analyze performance on each essential learning standard and set improvement goals for engagement and action. As an additional benefit, Elizabeth's team also received a fully populated spreadsheet of student data by standard to use for team analysis of overall student performance by standard for the unit.

TEACHER *Reflection*

How do your students make sense of their end-of-unit assessment results based on feedback received for each essential learning standard? For which standards have they demonstrated evidence of learning? Which have they not learned *yet*? What will you expect them to do if they are not yet proficient on certain standards?

_____

_____

_____

_____

_____

_____

_____

First graders might say, "I can compare numbers, but I can't write my numbers to 120 by myself yet." Similarly, a struggling high school student might say, "I can find the zeroes of a polynomial function, but I still need to learn how to determine asymptotes for rational functions."

These types of statements provide hope to students because they realize that they have learned some concepts and skills, and then they only need to focus on a limited number remaining, in order to demonstrate proficiency with understanding, application, and procedural fluency. A student is not "bad" at mathematics but rather has a few things yet to learn well. He or she needs additional practice.

The story from Bill Barnes reveals how student goal setting and action in response to your assessment feedback might have a few road bumps.

Consider how Elizabeth Kunstman's teacher team used Google Sheets to record assessment scores and engage their students in the goal-setting process. How can you and your colleagues take action after every end-of-unit mathematics assessment? Consider the following two teacher team actions.

1. Create a student goal-setting reflection process to identify errors, and use the assessment results to form a plan of action.

2. Create a process for students to act on their plan and take action (and allow them to improve their score for each essential learning standard on the end-of-unit assessment).

Simply assigning a summative grade to an assessment is the least useful strategy to motivate students to further their understanding (Kanold, Briars, Asturias, Foster, & Gale, 2013). Wiliam (2011) asserts, "As soon as students get a grade on a test, the learning stops" (p. 146).

Learning should not stop when your collaborative team provides meaningful assessment feedback to students; you might be surprised to know that their score or grade is the *least important* aspect of the assessment. Students must take action on their feedback from the unit assessment and use it as a formative learning opportunity instead of a *test once, then done* approach. Your feedback allows students to reflect on their strengths and challenges from the assessment.

When you pass the test back in class, give students time to self-assess their results, and make a plan of action to retool their learning. Wiliam (2018) describes that effective feedback should:

> *Cause thinking* by creating desirable difficulties. Feedback should be focused; it should relate to the learning goals that have been shared with the students; and it should be more work for the recipient than the donor. Indeed, the whole purpose of feedback should be to increase the extent to which students are owners of their own learning. (p. 153)

To ensure student ownership of learning with continued action, your team's feedback to students should be more than a score. Figure 4.4 is a sample student self-assessment form for the grade 4 unit assessment on fractions in figure 3.3 (page 43). This self-assessment form also aims to be a communication tool sent home for parents to sign as an acknowledgment of their child's progress in learning mathematics. Notice how much more information this reflection provides than an overall percentage or score on the assessment. This form takes advantage of using the intentional team design of writing the common mathematics assessment by essential learning standard.

Figure 4.5 shows another sample form for student self-reflection to use after returning the sample high school algebra 1/integrated mathematics I unit test (figure 3.4, page 47) to students.

The intent of students using the self-reflection form in class when you pass back the mathematics unit assessment (test) is to help them self-assess their performance and build responsibility for their own learning. From the reflection, students can set learning actions around those concepts and skills not yet learned.

Your team can also determine how all students will work to learn those standards during the next unit or through the interventions discussed in the next chapter.

Note that figures 4.4 and 4.5 show student self-reflections that include an analysis of both what students do know and what they don't yet know. This is intentional and of critical importance.

One of the critical factors known to promote access and equity is teaching practices that build on the *strengths* of students (Fernandes, Crespo, & Civil, 2017). As such, you and your colleagues must be careful to avoid a deficit orientation with respect to your students' knowledge and skills.

Too often, teachers on a collaborative team focus almost exclusively on what students do not know. However, to support students in recognizing they are

---

Personal Story 66 **SARAH SCHUHL**

I learned quickly that when providing more focused student reflections with error analysis, students viewed every missed question on an assessment as a simple mistake and did not always recognize conceptual errors in reasoning. In fact, after designing a complex reflection sheet and distributing it to my students in calculus and algebra 1 on their respective assessments, I realized that students could not accurately learn from the reflection sheet because they did not understand *how* to reflect in mathematics.

As such, I had to step back and teach self-reflection through sharing samples of positive student work on the unit assessment, and ask my students to identify the strengths and areas to improve from those anchor papers. Once students were able to see the benefits of exploring mistakes and they were better able to identify strong and weak work, they began to also better self-reflect.

Name: _____ Date: _____

Grade 4 Fractions Unit
Student Reflection: Unit Assessment—What Have I Learned? What Have I Not Learned Yet?

Essential Learning Standards	Test Questions	Score	How I Did (Circle one.)	
1. I can explain why fractions are equivalent and create equivalent fractions.	1–4	____ out of 8	I got it! 6–8	I'm still learning it. 1–5
2. I can compare two fractions and explain my thinking.	5–9	____ out of 12	I got it! 9–12	I'm still learning it. 1–8
3. I can add and subtract fractions and show my thinking.	10–14	____ out of 15	I got it! 11–15	I'm still learning it. 1–10
4. I can multiply a fraction by a whole number and explain my thinking.	15–17	____ out of 7	I got it! 5–7	I'm still learning it. 1–4
5. I can solve word problems involving fractions.	18–20	____ out of 6	I got it! 4–6	I'm still learning it. 1–3
Learning standards I know and can do:			**Learning standards I am still learning:**	

Figure 4.4: Sample student end-of-unit self-assessment—grade 4 fractions.

Name: _____ Date: _____

Algebra 1/Integrated Mathematics I Linear Functions Unit
Use the following to reflect on the questions: What Have I Learned? What Have I Not Learned Yet?

Essential Learning Standards	Test Questions	Score	Percentage
1. I can graph linear functions and interpret their key features.	1–5	/13	
2. I can calculate and explain the average rate of change (slope) of a linear function.	6–8	/6	
3. I can recognize situations with a constant rate of change and interpret the parameters of the function related to the context.	9–12	/11	
4. I can construct linear functions given a graph, description, or two ordered pairs.	13–17	/18	
5. I can graph piecewise functions, including step and absolute value functions.	18–21	/16	

My strengths (the essential learning standards I learned):

My areas for growth (the essential learning standards I am still learning):

My learning goal and plan:

Figure 4.5: Sample student end-of-unit self-assessment—High school linear functions unit.

mathematically capable, you must also acknowledge what students do know (Jilk & Erickson, 2017). By acknowledging students' strengths, students have the opportunity to develop a positive identity of themselves as learners of mathematics (Aguirre, Mayfield-Ingram, & Martin, 2013).

Collaboration with peers and ownership of the learning pathway engage students more deeply in the learning process and provide evidence for each student that effective effort is the route to mathematical understanding and success.

Formative reflection is not an easy process for students, however, and must be taught, as Sarah described in her story on page 76. You and your colleagues will need to determine a systematic form and process to use after each end-of-unit mathematics assessment. This allows your students to learn a process for analyzing their work and interpreting your feedback. While it will take time to teach students this critical reflection skill, over time, students learn to more quickly and accurately identify what they have learned and not learned yet, as well as understand the effort required to continue learning due to any errors made on the assessments' mathematical tasks.

At every grade level from high school through kindergarten, students need to learn how to accurately self-reflect on their progress toward learning essential standards and set goals. And, at every grade level, students *can* do this work and learn from the results.

TEACHER *Reflection*

How do your students learn from their errors on end-of-unit assessments? How do they equate that learning to an understanding that they are learning an essential learning standard, rather than just simply learning how to do a specific task?

_____

_____

_____

_____

_____

_____

Your team may also decide to use a more complex data-tracking chart like the one in figure 4.6 that the algebra 1 teacher team in the Norwalk Community School District, Iowa, used. Students record their current proficiency on each essential learning standard throughout a semester. This kind of recording form can be part of a data notebook and something you use for student-led conferences along with the common assessments and self-reflection forms.

At the DuFour Award–winning Mason Crest Elementary School in Annandale, Virginia, kindergarten students have essential learning standards on strips of paper attached to a large ring. One paper, for example, might show the essential learning standard: *I can count to 100*. Underneath the essential learning standard are benchmarks such as 1–10, 11–20, 21–50, and 51–100, each written below a star. As students meet each benchmark on the way to becoming proficient with the essential learning standard, they hole-punch or color the corresponding star. Over the course of the year, students record their continued learning for each essential learning standard (J. Deinhart, personal communication, June 1, 2017).

With your collaborative team focused on student learning, it is important students understand that their end-of-unit reflection will in some way require them to re-engage in learning standards that fall in the *not yet* category.

Additionally, you and your colleagues need to consider how students will demonstrate learning the mathematics standards by making every unit assessment more formative in nature. This requires a team response and a student response. It also affects your teams' grading decisions, discussed in more detail in *Mathematics Homework and Grading in a PLC at Work* in this series.

As one final example, consider figure 4.7 (page 81). This is a sample from Oak View Middle School in Andover, Minnesota. Students reflect on their learning using evidence from an assessment and then use the FAST feedback idea to commit students to a learning plan (contract) that includes correcting assessment tasks, often collaboratively.

Alternately, some schools simply ask students to sign a contract to retake an end-of-unit assessment that requires the student to articulate which learning

continued →

Name: _____   Period: _____

Algebra 1 First Semester Proficiency Graph

4.0	3.5	3.0	2.5	2.0	1.5	1.0	0.5	0.0			
		Proficient							★**Write Equations of Lines Assessment**		Write Equations of Lines
									Lines of Best Fit CFU		
									Parallel and Perpendicular Lines CFU		
									Write Equations of Lines CFU		
									★Graph Equations Assessment		Graph Equations
									Graph Equations in Standard Form CFU		
									Graph Equations in $y = mx + b$ CFU		
									★**Slope Assessment**		Slope
									Slope (Rate of Change) CFU		
									★**Interpret Functions Assessment**		Interpret Functions
									Arithmetic Sequences CFU		
									Function Notation CFU		
									Functions, Domain, and Range CFU		
									★**Create and Graph Functions Assessment**		Create and Graph Functions
									Write and Graph Functions CFU		
									Write and Graph Function Patterns CFU		
									★**Solve Inequalities Assessment**		Inequalities
									Absolute Value Equations and Inequalities CFU		
									Compound Inequalities CFU		
									Solve Multistep Inequalities CFU		
									★ **Solve Proportions Assessment**		Equations
									Solve Proportions CFU		
									★ **Solve Equations Assessment**		
									Solve Equations CFU (2.2–2.5)		
									Solve Variables on Both Sides CFU		
									Solve Multistep Equations CFU*		

Learning Targets

*CFU = Check for understanding

Color key: ☐ First attempt   ☐ Second attempt   ☐ Third attempt

Figure 4.6: Student progress-tracking chart from algebra 1 in Norwalk Community School District.

Name: _____

Period: _____

Algebra 1 Second Semester Proficiency Graph

Learning Targets		
★ Solve Quadratics Assessment		Solve Quadratics
Solve Quadratics CFU		
Solve Quadratics CFU		
★ Graphing Quadratics Assessment		Graphing Quadratics
Graphing Quadratics CFU		
★ Factoring Polynomials Assessment		Factoring Polynomials
Factoring Polynomials a > 1, Grouping CFU		
Factoring Polynomials a = 1 CFU		
★ Operations of Polynomials Assessment		Operations of Polynomials
Multiply Polynomials CFU		
Add and Subtract Polynomials CFU		
★ Exponential Functions Assessment		Exponential Functions
Geometric Sequences CFU		
Exponential Growth and Decay CFU		
★ Exponent Properties Assessment		Exponential Properties
Rational Exponents and Radicals CFU		
Exponent Properties CFU		
★ Systems of Linear Inequalities Assessment		Systems of Linear Inequalities
Systems of Linear Inequalities CFU		
★ Solve Systems of Equations Assessment		Systems of Equations
Solve Systems by Elimination CFU		
Solve Systems by Substitution CFU		
Solve Systems by Graphing CFU		

Proficiency scale (y-axis): 4.0, 3.5, 3.0, 2.5, 2.0, 1.5, 1.0, 0.5, 0.0

Proficient line at 3.0

Color key: ☐ First attempt   ☐ Second attempt   ☐ Third attempt

Expression Test Reflection

Learning Targets for Grade 8, Unit 4, Part A:

1. I can use the distributive property and combine like terms to simplify algebraic expressions.

2. I can evaluate algebraic expressions.

3. I can translate from a real-world situation to an algebraic expression.

4. I can use formulas with multiple variables.

Retake Option

If a learning target score on your assessment is below 70 percent, you are **required** to retake the assessment to improve your score on the learning target. You must complete the following problems listed for the learning targets for which you scored below 70 percent. This may mean you retake all four learning targets or just one of them. Turn them in when you come to your scheduled re-take time.

Learning target 1: Ask your teacher for extra practice and submit by _____ date.

Learning target 2: Ask your teacher for extra practice and submit by _____ date.

Learning target 3: Pages 1–9, 12, and 20

Learning target 4: Pages 1–8, 304

Learning Contract

In order to retake parts of this test, you must complete the following items.

☐ Complete and staple the unit 4: part A test corrections to the original test (other side).

☐ Complete and turn in all practice from during this unit.

☐ Complete and turn in the study guide for the test.

☐ Complete, grade, and turn in re-take practice problems (by standard) for the learning targets for retake.

What other actions will you take to learn each standard from this expressions unit assessment?

_____   _____
Student Signature   Date

_____   _____
Parent Signature   Date

Figure 4.7: Example of grade 8 student contract and test correction form.   continued →

Name: _____

Test Correction Form

Attach original test to this form.

Number	Reason I Got This Wrong	Correct Answer	Reason My New Answer Is Correct

Source: © 2014 by Oak View Middle School. Used with permission.

targets he or she learned and which he or she still needs to learn with an accompanying learning plan and agreed-on date by which the learning will occur and the test will be taken.

The intent of students analyzing their performance on the end-of-unit assessment is to help each student build responsibility for his or her own learning. While each student takes ownership for his or her individual progress toward each of the essential learning standards, students may still work together to meet those standards. Students can work together when your team provides equitable feedback to students, regardless of which teacher they are assigned, on your common end-of-unit assessments.

Student-to-student feedback also is a vital component of the entire educative process, and goal setting at the end of each unit is part of student collaboration. As students compare and contrast their work with one another, they see what they learned well and what they still need to learn. Such collaboration with peers and ownership of learning engage students more deeply in the learning process and provide evidence for each student that effective effort leads to mathematical understanding and success.

TEAM RECOMMENDATION

Student Action on Feedback From the Common End-of-Unit Assessment

- Develop a team systematic form for students to use as a self-reflection tool after each end-of-unit assessment. Be sure students identify what they have learned as well as what they have not learned *yet*.

- Teach students to analyze their work and accurately self-reflect using the assessment feedback.

- Require students to collaborate, take action, and learn from their errors.

Requiring students to re-engage in learning essential standards and learn from your feedback, as well as feedback from peers, is generated and supported through the opportunities for intervention that your team and school provide. In the next chapter, you will examine the criteria of highly effective mathematics intervention programs.

TEACHER *Reflection*

How do students track their learning progress over time or after each end-of-unit assessment for your grade level or course?

What are some ideas you are considering to strengthen student reflection and action after each end-of-unit assessment, which you and your colleagues can employ?

CHAPTER 5

Team Response to Student Learning Using Tier 2 Mathematics Intervention Criteria

As long as interventions are viewed as an appendage to a school's traditional instructional program, instead of an integral part of a school's collaborative efforts to ensure all students succeed, interventions will continue to be ineffective.

—Richard DuFour

Not all students learn the same way or at the same rate, as evidenced by the student work on your end-of-unit common mathematics assessments. When students have not yet learned essential standards, the collaborative teacher team response in a PLC culture is not to stop and reteach all students. Such an action would be done at the expense of students learning the remaining essential standards that are part of the guaranteed and viable curriculum for your grade level or mathematics course.

Instead, the PLC response is to work together to design a quality mathematics intervention program that provides additional time and support to students needing to still learn grade-level or course-based essential learning standards, while continuing to teach the essential learning standards in the next unit. Such a mandate requires a team effort to ensure every student is learning from an intentional mathematics intervention. Quality systematic interventions are not pull-out programs or initial differentiated instruction variety, but rather a value-added model of continued instruction.

There are generally three tiers of mathematics intervention, which are included in any response to intervention (RTI) system of supports. RTI provides interventions for students as a whole group (level 1), in small identified groups (level 2), and as individuals (level 3). (For more information on the three tiers of RTI, go to www.allthingsplc.info/blog/view/335 /connecting-plcs-and-rti to read the blog of RTI expert Mike Mattos.)

It is possible that as you and your colleagues embrace new teaching strategies and use high-quality instructional practices, as described in *Mathematics Instruction and Tasks in a PLC at Work* from this series, it will help you limit the number of students who require the teacher team–developed, Tier 2–type mathematics interventions described in this chapter.

For the purposes of this book related to your teams' common assessments and interventions, the focus is on Tier 2–type interventions. The intent of your mathematics interventions should be to provide students with additional time and support needed to learn your grade-level or course-based essential learning standards.

There are two important observations from Mona Toncheff's story on page 86. First, the teachers at Field Elementary asked students to self-identify their areas of strength and weakness as part of the WIN process. Second, the teacher team established a viable Tier 2 response that allowed students to take action on their initial assessment performance and receive help and support from a wide range of faculty and staff.

Personal Story **MONA TONCHEFF**

When teams analyze their common assessments, the biggest hurdle to overcome is to plan, *now what?* When I was working with K–5 mathematics teams at Field Elementary in Louisville, Kentucky, the faculty and staff worked diligently to collectively answer the *now what?* question.

The collaborative teams met every Friday to review the common assessment data from the week and identified students in each grade level who were meeting, exceeding, or struggling with the essential standards. They decided their students would create a plan for re-engagement or enrichment during their designated WIN (**W**hat **I** **N**eed) time.

When the grade-level teacher team started this process, the teachers were reluctant to share their students' information during WIN time. The first year, some teachers would keep their students in their classrooms and try to differentiate mathematics instruction using small student groups. However, demonstrations of student learning were not improving. Students with learning differentials during the second week were still struggling with the same concepts in the twelfth week.

By midyear, the teacher teams were more willing to discuss instructional practices that were most effective on the agreed-on essential learning standards. They started rotating students to different teachers during WIN time to better target specific instructional practices or student strategies that would move students forward by standard. Student progress was continually monitored, and student movement was fluid and flexible during WIN time.

By the second year, the entire staff was also part of the solution. Support staff were assigned to designated groups or teams during WIN time. They also were part of the team and participated during team time to understand how to address the learning needs and support the learning outcomes for each student. The common assessment evidence drove their WIN time schedule, and teams were changing the status quo of student learning to address students' diverse learning needs. See AllThingsPLC's (n.d.) "Evidence of Effectiveness: Field Elementary" (www.allthingsplc.info/evidence/details/id,998) to read more about the school's success.

The questions in figure 5.1 aim to help your team understand one another's perspectives related to your systematic interventions as a team. In other words, how does your team respond when common assessment data reveal that some students have learned the essential learning standards and others have not? Your professional response as individual teachers and as a team will reveal your current beliefs about the need for interventions and the plan for making them effective.

How do you know if your intervention programs are high quality? The five criteria described in figure 5.2 (page 88) provide a mathematics *intervention quality* evaluation tool your collaborative team can use to evaluate the quality of your current mathematics intervention program as well as to strengthen your mathematics intervention practices in each grade level or course.

TEACHER *Reflection*

When an end-of-unit mathematics assessment shows that students have not learned essential learning standards, what is *your* response?

What is your *teacher team's* response?

What do you require all students in the grade level or course to do?

_____

_____

_____

_____

_____

Directions: Use the following prompts to guide team discussion of your current team intervention practices.

Purpose of a team intervention:

1. Why do you need a team response when designing interventions for essential learning standards?

Plan for team intervention:

2. How do you identify students in need of intervention?

3. How do you determine the targeted content and skills to address through Tier 2 interventions?

4. Who provides the Tier 2 interventions for students in your class or across your team? When? How do you and your colleagues share the responsibility of interventions for students across your grade level or course?

5. How are your interventions a just-in-time required opportunity for growth in learning?

6. How do students move in and out of interventions? How often?

Effectiveness of interventions:

7. How do you know if your interventions are effective?

8. How do students know their time spent learning in an intervention is effective?

Figure 5.1: Team discussion tool—Intervention practices.

*Visit **go.SolutionTree.com/MathematicsatWork** for a free reproducible version of this figure.*

Mathematics Program Tier 2 Criteria	Description of Level 1	Requirements of the Indicator Are Not Present	Limited Requirements of This Indicator Are Present	Substantially Meets the Requirements of the Indicator	Fully Achieves the Requirements of the Indicator	Description of Level 4
Systematic and required	Teachers provide optional opportunities, often before or after school, for students to get individual or small-group help.	1	2	3	4	Team requires students to learn if they are not yet proficient with essential learning standards and provides systematic structures during the school day to ensure student learning. Team commonly plans for continued learning in the next mathematics unit, as needed.
Targeted by essential learning standard	Teachers provide individual assistance as needed or gather a group of struggling students that a sole screener or diagnostic assessment identifies rather than a skill identified from common unit assessments.	1	2	3	4	Team designs and administers common unit mathematics assessments to analyze and collectively respond to the data to identify specific students needing targeted intervention by each essential learning standard.
Fluid and flexible	Teachers place students in an intervention where they remain due to results from a diagnostic assessment as a response to mandated services.	1	2	3	4	Team regularly analyzes student data from common unit mathematics assessments and allows students to move in and out of the required additional time and support in learning.
Just in time	Students get help before the end of a grading period or when they request help.	1	2	3	4	Students get real-time feedback, and the team plans its interventions for the essential learning standards within a mathematics unit or at the start of the next unit using current common assessment results.
Proven to show evidence of student learning	Teachers provide intervention to students and hope students learn. Students may be receiving intervention support from a person not highly qualified in mathematics. There is little evidence that the intervention is helping students learn the essential standards for the grade level or course.	1	2	3	4	Team monitors student progress within the intervention or within class to see if the intervention is effective and results in student demonstrations of proficiency learning for each essential standard. Team members continue monitoring instructional strategies that impact learning and re-engagement. Each adult implementing the team-designed mathematics intervention is the best-qualified person for the role.

Figure 5.2: Mathematics intervention program evaluation tool.

Use the tool in figure 5.2 to rate and evaluate the quality of your response to intervention and learning evidenced on a recent common end-of-unit assessment. How do your current mathematics intervention programs score? You should expect to develop an intervention program that scores 4s in all five criteria. The following section explains each of the five criteria in more detail.

TEACHER *Reflection*

Which of the five criteria for a high-quality mathematics intervention program is currently part of your collaborative team practice?

What do you need to do to strengthen your mathematics intervention program?

_____

_____

_____

_____

_____

Each of the five criteria is needed for continued student learning when a student has not yet learned an essential learning standard, as evidenced on the end-of-unit common mathematics assessment. The challenge is to create an effective system that allows students to engage in the intervention, while simultaneously practicing the essential learning standards for the next unit.

As described in figure 5.2, there are five criteria your team must consider when building your mathematics intervention programs.

1. Systematic and required

2. Targeted by essential learning standard

3. Fluid and flexible

4. Just in time

5. Proven to show evidence of student learning

Systematic and Required

When your team plans for additional time and support for students who still need to learn essential learning standards, time is often an issue. And until you address this issue, you will struggle to have a strong

Tier 2 program that is both systematic and required. Your team can solve some of these time issues, but others might require a coordinated effort with your leadership team.

To be systematic, your mathematics intervention program needs a consistent structure within the school day as part of the school system. To be required means that the students your team identifies as needing intervention cannot opt out of the additional learning.

You should recognize that some students are not learning the essential standards because of a lack of effort and choose failure. As Timothy Kanold indicates in his story on page 90, those students are choosing to not mow the lawn. Allowing such irresponsibility does not teach your students to become more responsible.

Thus, a required student action to learn is vital for a strong mathematics team intervention response. Students do not get the choice of failure, nor do they get to avoid re-engaging in learning the essential learning standards.

Think about the current intervention opportunities for students at your school. Now, ask yourself, how many of the interventions does your school require? Do students who need the intervention attend or participate in the intervention?

How do your team and school embrace the PLC culture to make it easier to pass than to fail? A required response means the intervention happens during the time students must be on campus and your team ensures students attend and participate in the learning opportunity. As you work together to determine appropriate interventions, you should also ensure each academic intervention is not happenstance.

Your collaborative team's response to student learning should address the following questions.

- Is the intervention needed on an individual basis, or does the gap exist in all students or particular subpopulations?

- Which essential learning standards do students still need to learn? Which specific mathematics concepts, skills, or applications do they still need to learn?

- Is the intervention tailored to meet the needs of a specific subpopulation of students?

- For English learners (ELs), what research-affirmed best practices for content literacy are you using?

- For inclusion students, what accommodations or modifications are you implementing to differentiate learning?

- What responsive teaching do you need for all students to be successful?

- What manipulatives or technologies should you use to develop greater conceptual understanding, application, or procedural fluency for each standard?

As you work to answer these questions, use the template in figure 5.3 to help you organize your thinking and create a team plan of mathematics intervention your students can access.

Depending on the number of students needing intervention with an essential learning standard, your team might also determine a plan to weave a required intervention into the instructional time it allocates to the next unit. The intervention might be:

- A lesson or series of lessons the team agrees to implement

- A flex day the team uses to share students, with each team member addressing a particular skill that students need to learn

- A concept intentionally knitted into specific and agreed-on lessons for the next unit (for example, continue working on solving word problems)

Overall, you can strengthen a Tier 2 mathematics program when it is a normal part of the school day or

week and when you require students to re-engage in learning across the grade-level or course-based team.

What Is the Mathematics Intervention?	Which Students Need Intervention?	How Often Will Students Get Support?	How Will Students Engage in Intervention?

Figure 5.3: Team discussion tool—Mathematics intervention planning tool.

Visit go.SolutionTree.com/MathematicsatWork for a free reproducible version of this figure.

TEACHER *Reflection*

How will you and your colleagues work with your school administrators to design a mathematics intervention program that is systematic and required?

_____

_____

_____

_____

_____

_____

Targeted by Essential Learning Standard

After your team analyzes the data from a common unit assessment, you will determine specifically which students are proficient with each essential learning standard and which are not. (See *Mathematics Coaching and Collaboration in a PLC at Work* in this series for a common unit-assessment data-analysis protocol.) You

also need to identify trends in student reasoning that must be corrected.

Your team can then create an intervention plan to determine which students need to re-engage in learning each essential learning standard and how to most effectively target that learning. Ideally, through self-reflection, students might be able to tell you which essential learning standard they most need to work on during the intervention.

Mathematics interventions are not as productive or consistent for students when teachers make individual decisions about how to respond to learning. Nor is the learning equitable for all students. The interventions must reflect your teacher *team* response to student learning so each student in the grade level or course receives the same access to learning the team-developed expected outcomes. This is an equity issue for you and your colleagues.

If your systematic mathematics intervention time occurs during a portion of the day when each student must be assigned a location (for example, recess, assembly schedule, or walk-to-mathematics), your team will also have to plan for targeted extensions for students who have already learned the essential learning standards in the unit. Determine a targeted extension that does not require students to peer-tutor but rather challenges each student to continue on in his or her learning journey.

You and your colleagues should create quality mathematics interventions focused on a targeted essential learning standard or skill. In a PLC culture, your team works together to determine the most effective intervention plans to help each student respond to the feedback they receive from your common unit mathematics assessments.

TEACHER *Reflection*

How will your mathematics intervention program group students by targeted skill so students can re-engage in learning a specific part of an essential learning standard?

_____

_____

_____

_____

_____

_____

_____

_____

_____

_____

_____

_____

_____

_____

_____

_____

_____

In addition to being systematic and required and targeted to specific essential learning standards, an effective mathematics intervention program must also meet the learning needs of students in a fluid and flexible way.

Fluid and Flexible

A third criterion for a quality mathematics intervention program is that students fluidly and flexibly move into and out of needing additional learning supports once your team establishes the mathematics interventions you plan to provide students at the end of every unit. There are two critically important questions you must ask and answer based on student assessment results.

1. How do we provide access for students to each mathematics intervention program?

2. How do we help students exit the mathematics intervention program once they exhibit proficiency on the targeted deficient mathematics standards?

Your expectations for mathematics intervention should not *permanently* place your students in a systematic and required intervention. Once students have demonstrated mastery of the expected mathematics learning standards, you should remove the intervention. Moving forward, the daily mathematics lesson time or class should be sufficient.

For example, if you are working with elementary students in a walk-to-mathematics program or high school students in a mathematics tutoring room, you may be able to target students after a unit assessment for a two- to three-week period of time. Once your students have learned the essential learning standards, you can remove them from additional support either at the end of the designated time period or before.

If, however, part of your Tier 2 mathematics intervention program in middle school or high school includes an additional class period of support, moving students into and out of the support is not as fluid or flexible due to other constraints on a student's schedule. This will be something you and your colleagues must address, as students need to know the mathematics intervention is not forever and can be removed from their schedule once they reach proficiency.

Moving students into and out of your intervention program is essential if students are going to have additional time and support to learn essential learning standards when that learning is most critical. Yet, all this effort will be for naught unless you can re-engage your students in areas of deficient learning soon after any common unit assessment.

Personal Story SARAH SCHUHL

A second-grade student stood to explain a subtraction problem that required regrouping to the class. Later, the student's teacher asked if I had noticed the student's thorough explanation. She marveled at his ability to subtract and exclaimed, "Can you believe it? He's one of my Tier 2 students!"

I simply replied, "Well, he's not Tier 2 related to subtraction."

In fact, no one is a Tier 2 or Tier 3 student. These are not labels of learning but rather structures to make sure students are getting the additional supports they need to learn.

TEACHER *Reflection*

How often will you move students into and out of your Tier 2 mathematics interventions and supports?

How will you structure your intervention program to make it fluid and flexible?

How much choice do students have for the nature of the intervention and support they receive?

_____

_____

_____

_____

_____

_____

_____

Sarah's story illustrates an important aspect of student ownership for the type of structured intervention you offer students. Do they have various options for the type of help they receive? You can find examples of these options in chapter 6 (page 97).

Your students should know that the purpose of the intervention is to help them demonstrate learning the mathematics standards—*just in time.*

Just in Time

As previously described, a Tier 2 mathematics intervention program should be systematic and required, targeted by essential learning standard, and fluid and flexible. Additionally, to be effective, the program must provide students with just-in-time feedback in order to grow their learning. This means the intervention focuses on essential learning standards students are currently learning or have just learned in the previous mathematics unit.

When a Tier 2 intervention program only uses district benchmark assessments or universal screeners to determine who should be in the intervention program, several problems for student learning may occur. Universal screeners and benchmark assessments seldom provide feedback specifying targeted skills or standards that students need to learn (not targeted by essential learning standard). Additionally, educators don't give universal screeners and benchmark assessments frequently enough to allow for a fluid and flexible program. Nor are universal screeners and benchmark assessments able to give just-in-time feedback to students.

While they do provide additional data and can be informative, universal screeners and district benchmark assessments alone do not give your team the most effective data for a quality Tier 2 program, nor are those mathematics assessments generally used for formative student learning purposes. It is only through your *team-created common assessments that you give students just-in-time feedback and allow your Tier 2 program to re-engage students in learning necessary for success in that moment of time.*

When Tier 2 interventions are not tied to current or recent essential learning standards, students sometimes miss the purpose of re-engagement in learning because it lacks connections. Using a computer program alone for interventions when not assigning targeted modules for students to complete can contribute to students getting feedback, but not feedback related in a timely way to anything they are currently learning. This is problematic. Too often, this causes student confusion and teacher frustration because students do not necessarily grow in their learning despite the time spent in intervention.

TEACHER *Reflection*

How does your current mathematics intervention provide just-in-time learning of the essential learning standards students have not learned yet but have addressed in the current or previous unit?

_____

_____

_____

_____

_____

_____

A just-in-time systematic Tier 2 mathematics intervention program does not ask students to get help weeks after learning but rather, *right away.* Finally, your team will only know if a Tier 2 mathematics intervention program is successful if there is evidence of student learning.

Proven to Show Evidence of Student Learning

Once your team determines a targeted intervention plan, you must also determine if the intervention was effective. Did it actually improve student learning in the targeted essential learning standard or skill? How do you know?

When determining effectiveness, further demonstrations of student learning are most important. Your team members will need to ask how the instructional strategies used and the teacher who employed the intervention impacted that student learning.

The National Council of Supervisors of Mathematics (NCSM) advocates that quality Tier 2 mathematics intervention learning opportunities require teachers who are most skilled in teaching mathematics (NCSM, 2015). The most skilled teachers for Tier 2 interventions are usually the professional educators teaching that grade level or course.

In a PLC culture, your team owns Tier 2 interventions for students and does not think of these interventions as places to where or persons to whom they send students who have not yet learned in hopes that someone else fixes their learning. Rather, it is a teacher- and team-owned program and process for greater effectiveness.

Additionally, mathematics support and interventions in Tier 2 should look significantly different from the students' initial learning experience (NCSM, 2015). For example, students in first grade who have not learned addition by using Unifix cubes or ten-frames may need an alternate strategy to build conceptual understanding.

In later grades, a Tier 2 intervention focusing only on re-teaching an algorithm does not often produce more learning. Developing an understanding for the algorithm and connecting it to previous learning may better grow student understanding and application.

When looking at your common assessment results, your team should determine best instructional practices that result in student learning and use that information to strengthen both Tier 1 instruction and Tier 2 interventions. During this time, your team may want to determine best uses for manipulatives and other technologies or digital resources. How do you give students access to these resources, and how do students use them for explorations to deepen conceptual understanding or to visualize or make sense of real-life applications and modeling?

As your team designs Tier 2 intervention experiences tied to essential learning standards, consider who is the best person to lead each group of students and how students should be re-engaged in learning the content. Then, consider how you will know whether or not students actually increased their learning *because* they engaged in the intervention.

You will want to use short, team-created common assessments during the mathematics intervention to see if students have learned the essential mathematics standard for which they are receiving additional support. This intervention assessment should provide opportunities for students to retake all or part of a during-the-unit or end-of-unit exam. Knowing whether or not an instructional strategy improved learning and whether or not students have grown in their learning due to the intervention you provided is critical to an effective Tier 2 mathematics intervention program.

Who provides the Tier 2 mathematics intervention your students need?

How do you decide and what data do you examine if the intervention is effective for closing gaps in student learning of the essential learning standards?

_____

_____

_____

_____

_____

_____

_____

Mathematics intervention is about creating equity and access to meaningful mathematics for your students. Tier 2 intervention *does not* mean placing all students with learning difficulties in the same classroom and pulling them out of core instruction or high expectations for their work. Tier 2 *is not* leveling students at the elementary or secondary level and giving them the same amount of time to learn, but as a lower-ability group.

NCTM (2014) explains, "The practice of isolating low-achieving students in low-level or slower paced mathematics groups should be eliminated" (p. 64). The Tier 2 intervention should provide *additional* time and support to learn the essential standards.

Kanold and Larson (2012) advocate that efficacious intervention programs should:

- be mandatory, not optional (i.e., scheduled during the school day whenever possible)
- be based on constant monitoring of students' progress, as determined from the results of formative and summative assessment, ensuring that students get support as quickly as possible;
- attend to conceptual understanding as well as procedural fluency; and
- allow for flexible movement in and out of the intervention as students need it (p. 66)

Regardless of the structures you have in place to provide additional time and support, your collaborative team members will need to navigate the complexities of the school day and collectively respond to the evidence of student learning by requiring students to participate in coordinated intervention efforts. The interventions you provide should not be a student choice. As soon as a mathematics unit ends, students who are not able to demonstrate learning of the standards should be immediately required to engage in mathematics intervention and support. Thus, you should design your feedback for assignments, assessments, and classwork to allow students to create an action plan.

Imagine, if you will, a K–12 student who is struggling in mathematics. Is he or she allowed to give up, stop trying, or take an "easy" zero? Or does he or she know your team has his or her back? Does the student know there are guaranteed supports in place to require him or her to succeed because his or her teachers believe the student can learn at high levels?

A quality, team-created Tier 2 mathematics intervention program erases the inequities team members create when they design isolated interventions and fail to be transparent with one another. You should make your end-of-unit intervention response a priority for the work of your team.

TEAM RECOMMENDATION

Create a Tier 2 Mathematics Intervention Program

Use common assessments to create a team response to student learning that is:

- Systematic and required
- Targeted by essential learning standard
- Fluid and flexible
- Just in time
- Proven to show evidence of student learning

Determine how often your team will meet to plan for an effective Tier 2 intervention response and how you will hold one another accountable to your team intervention plan.

In chapter 4 (page 67), you examined calibrating common assessment scoring and providing students with consistent and accurate FAST feedback as part of a strong, formative assessment process. Students then use your FAST feedback to self-reflect and set goals for continued learning. Your team's robust Tier 2 mathematics intervention program provides the resources and support students need for that continued learning. Together, these ideas create a meaningful and productive use of your team-created common assessments. Together, you have created a more formative process for student learning.

Not surprisingly, you can make the formative assessment process stronger if your students do not wait until after the end-of-unit common assessment to self-reflect. Do you give students opportunities to proactively reflect on their progress *during* the unit? This continual reflection allows for additional learning as needed during a unit so students can more confidently demonstrate evidence of learning on the end-of-unit assessment.

CHAPTER 6

Student Action on Assessment Feedback During the Unit

When students engage in their learning and take ownership of the process,
they are able to set goals, evaluate their progress, and determine what course
of action to take to reach the highest level of achievement.

—*Sharon V. Kramer*

There is a final assessment and intervention for you and your colleagues to consider as part of your unit-by-unit work in mathematics. In addition to end-of-unit common formative assessments, students can also learn and benefit from shorter common assessments provided during a mathematics unit. Since each assessment closely aligns to one or two essential learning standards, students can use your feedback to determine whether they have learned the standard yet before the final end-of-unit assessment. It is important that students take action on and ownership of the assessment feedback they receive during the unit. They will then be able to engage in Tier 2 mathematics interventions before the unit ends as you help them respond to PLC critical question 3 (DuFour et al., 2016): How will we respond when some students do not learn (in this case, during the unit)? Thus, learning during the unit becomes more formative.

NCTM's (2014) *Principles to Actions* states:

> At the center of the assessment process is the student. An important goal of assessment should be to make students effective self-assessors, teaching them how to recognize the strengths and weaknesses of past performance and use them to improve their future work. (p. 95)

Think about who is working harder to close students' learning gaps—you or your students? How can you work together with students to encourage and increase their ownership of learning?

TEACHER *Reflection*

How are students able to reflect *during* a unit on their progress toward meeting an essential learning standard?

When students recognize that they still need to learn essential learning standards during a unit, what structures are in place for them to re-engage in that learning?

_____

_____

_____

_____

_____

_____

Your students can participate in their own learning in many ways, which allows them to track their own progress and set their own goals. Two strategies include student trackers and self-regulatory feedback.

Student Trackers

Ridgeview Middle School in Visalia, California, uses a tracker sheet to build a routine way of having students document their learning toward the essential standards in each unit (see figure 6.1) based on feedback it receives or generates during the unit. Students receive the tracker on cardstock at the start of the unit. At the top of the trackers, essential learning standards appear as student-friendly learning targets. They reference these targets throughout the entire unit, identify them for each common homework assignment (for more on this idea, see *Mathematics Homework and Grading in a PLC at Work* from this series), and write them on each common assessment within the unit as well as at the end of the unit for each associated group of questions.

The team has created additional learning opportunities for students, which students choose from on the second side of the tracker while they track their progress. (This is part of the team's Tier 2 mathematics intervention program based on their assessment performance; see figure 6.2, page 101.)

Students also continue to monitor their progress using the shading bars to the right of each learning target, and shade (or erase) based on evidence generated from warm-ups, classroom tasks, homework, exit slips, and common checks for understanding used during the unit.

Because students are clear about what they should be learning, they are able to articulate what they still need help with and can be part of the intervention process in a productive way. They can even work to learn the content with other students in the same course who may have different teachers, expanding their resources for feedback. Additionally, your students become more aware of which mathematics standards they most need to study and learn prior to the end-of-unit assessment.

At elementary grade levels, student trackers should have the essential learning standards written as student-friendly targets and may also have common homework (intermediate grades) or even a more detailed proficiency scale description with sample problems shown, so students better understand the meaning of each scale.

A kindergarten class, for example, used take-home folders for students. The teacher agreed to two mathematics and two ELA targets for a unit of time. She glued four Velcro strips on the inside of the folder (right side), and each Velcro strip had a Velcro circle for students to move as they learned the target. On the left

---

> **TEACHER** *Reflection*
>
> After you review the student tracker in figure 6.1, think about how your team could integrate a similar system of student ownership for the work of your mathematics class during the unit.
>
> _____
>
> _____
>
> _____
>
> _____
>
> _____
>
> _____
>
> _____
>
> _____

side of the folder, the teachers typed the learning targets so they could be changed as needed. The added benefit to this practice is parent knowledge about the most important essential standards students are working on.

Figure 6.2 shows an example of a third-grade tracker with a proficiency scale. Students check their progress as they move from one level to the next for each learning target and also stop at key points during the unit to identify their actions to continue learning.

You can modify for all grade levels any sample student trackers you use to build routines around student self-assessment and action during a unit. You should *require* students to use trackers and also to re-engage in learning. Due to your robust response when students are not learning, and your expectations for them to track their own progress, you signal to students that learning the essential standards for mathematics is of utmost importance.

One final example at the high school level includes students identifying the percentage earned by essential learning targets on each common mid-unit assessment or quiz as well as on the end-of-unit assessment (see figure 6.3, page 102). Students can determine if there is growth and identify their level of mastery after the unit assessment. After each quiz, students identify, if needed, how they will re-engage in learning the essential learning standard and identify time to take a recovery quiz prior to the end of the unit. The options for continued learning require the team to have a systematic Tier 2 response.

Name: _____ Period: _____ Date: _____

Student Tracker Unit 3: Expressions, Equations, and Inequalities

Learning targets:

1. I can simplify algebraic expressions.

Starting . . .	Getting there . . .	Got it!

2. I can write and solve an equation from a word problem.

Starting . . .	Getting there . . .	Got it!

3. I can write and solve an inequality from a word problem.

Starting . . .	Getting there . . .	Got it!

4. I can graph the solution to an inequality and explain how the solution answers the question.

Starting . . .	Getting there . . .	Got it!

Note: Teacher will use one to two warm-up questions that relate to each of the preceding learning targets to show student proficiency. Students must complete the following assignments to prepare for the warm-up questions.

Common Homework Assignments

Target Number	Lesson	Page Number	Problem Numbers	Assignment Status (Complete or Incomplete)

Vocabulary

Term	Description	Example
1. Expression versus equation		
2. Coefficient		
3. Constant		
4. Distributive property		
5. Inequality		
6. Solution to an equation versus an inequality		

Figure 6.1: Grade 7 student tracker—Self-assessment and action plan.

continued →

Test or Quiz Name	Testing Date	Original Test Score (Original score you earned when you took the test)	Correct and Reflect Score (3—Completed correctly 2—Incomplete and needs corrections 1—Never turned in)	Retake Score (Required if your score is less than 2.5)	Gradebook Score for This Test (Most recent score on the original test or the retake)

Check one:

☐ I am maintaining a proficient level for this class (3.0).   ☐ I need to put forth more effort in order to be proficient (3.0).

Action steps for improvement: Place a check mark in the shaded boxes under the tasks you will do to re-learn essential standards, as needed.

Test or Quiz Name	Morning (a.m.) Tutoring With Teacher (Available from 7:45–8:15)	Afternoon (p.m.) Tutoring With Teacher (Available from 3:15–3:45)	Tutoring With a Peer (Before school, lunch, or after school)	Completed Unit Study Guide	Homework Status (I have completed all my homework for this test.)	Correct and Reflect (I can do this on my own and finish before future due dates.)	After-School Intervention (I need to sign up for this every weekend.)

How Can I Raise My Grade?

Successful mathematics students:

- Come to class prepared (organized binder with completed homework, pencils, calculator, textbook, and paper) and are ready to learn every day
- Ask for help before they take an assessment, so there are no surprises
- Ask questions in class, do their very best on every assignment, and know where they still need extra practice
- Complete the correct and reflect in a timely manner if he or she earns a failing grade on any test. Retakes are available once the correct and reflect is done correctly.

What do you still need to work on in order to be successful this year?

Everyone can be successful!

Student signature: _____   Parent or guardian signature: _____

Name: _____ Date: _____

Student Tracker for Grade 3: Perimeter and Area Unit

In this unit, we are going to learn about area and perimeter. The learning targets with proficiency scales are below. Check each level you can do to show what you have learned so far and what you have not learned yet.

I can find the perimeter of polygons.

	This means I can:	What I Can Do	My Plan to Learn
4	Create two different polygons that have the same perimeter, and explain how I made them.		
3	Find the perimeter of a polygon if I know the side lengths or if I have to figure out the side lengths myself.		
2	Find the perimeter of a polygon when I am shown all the side lengths.		
1	Find the perimeter of a polygon with help.		

I can find the area of a rectangle.

	This means I can:	What I Can Do	My Plan to Learn
4	Find a missing side length on a rectangle if I know its area, and find two or more different rectangles that have the same area.		
3	Find the area of a rectangle using a formula and explain why the formula works.		
2	Find the area of a rectangle by making an array and counting the unit squares.		
1	Find the area of a rectangle by counting unit squares.		

I can find the area of a complex shape made of rectangles.

	This means I can:	What I Can Do	My Plan to Learn
4	Create two different complex shapes made of rectangles that have the same area, and show all the side lengths for each.		
3	Find the area of a shape made of rectangles if I know all the side lengths I need or if I have to find some of them first.		
2	Find the area of a complex shape made of rectangles when I am shown all the side lengths I need.		
1	Find the area of a complex shape made of rectangles when I am shown all the side lengths I need and have some help.		

Figure 6.2: Grade 3 student tracker—Self-assessment and action plan.

*Visit **go.SolutionTree.com/MathematicsatWork** for a free reproducible version of this figure.*

Name: _____   Date: _____

Student Tracker for High School: Geometry Right Triangles Unit Five

For each essential learning standard, record your points earned on the corresponding Points Quiz, and then decide how well you understand the learning standards at this time. For the final assessment, record how many points you earned for each essential learning standard, and then determine what kinds of mistakes you made and your target level of mastery for each learning standard.

For each learning target you are approaching, create an action plan of how you will learn the standards, and be prepared to take the recovery quiz (your second attempt at demonstrating mastery). Please remember to also write times and dates. The more specific you are about your plan, the more likely you are to stick to it. "Whenever" is not specific!

Standards	Learning Target (I can. . .)	Point Quiz — Right Triangles Readiness Check — How well do I know it? Exceed (90–100) — Points Earned	Percentage	Tutoring (Circle yes or no.)	Unit Test — Questions on Test	Level of Accuracy — Why did I not earn full credit? Proficient (70–89) — Points Earned	Percentage	Level of Mastery (in percentages) — Approaching (50–69)	Some Evidence (25–49)	No Evidence (0–24)
G-SRT.6	Explain the definitions for trigonometric ratios for acute angles in right triangles using similarity.	___ out of 8		Yes (0–69 percent) / No (70–100 percent)	1–3	___ out of 6				
G-SRT.7	Explain and use the relationship between the sine and cosine of complementary angles.	___ out of 8		Yes (0–69 percent) / No (70–100 percent)	4–7	___ out of 8				
G-SRT.8	Solve applied problems with right triangles using trigonometry and the Pythagorean theorem.	___ out of 15		Yes (0–69 percent) / No (70–100 percent)	8–12	___ out of 12				

Checkpoint quizzes are given to monitor learning and to address misconceptions with tutoring before the unit test.

Action Plan

Concept Mastery Options I Will Attempt Prior to Taking a Recovery Quiz
(Check all that apply, and write dates and times: morning, lunchtime, advisory, or after school.)

Standards	Learning Target (I can. . .)	Standard Recovery Tutoring With Teacher	Standard Recovery Tutoring in the Mathematics Learning Center	Standard Recovery Tutoring With Another Teacher or Student	Standard Recovery by Independently Completing Review	Standard Recovery by Rereading Notes	Standard Recovery by Online Program	Date of Recovery Quiz	Score
G-SRT.6	Explain the definitions for trigonometric ratios for acute angles in right triangles using similarity.								
G-SRT.7	Explain and use the relationship between the sine and cosine of complementary angles.								
G-SRT.8	Solve applied problems with right triangles using trigonometry and the Pythagorean theorem.								

Student signature: _____ Date: _____

Parent signature: _____ Date: _____

Figure 6.3: High school student tracker—Self-assessment and action plan.

Visit go.SolutionTree.com/MathematicsatWork for a free reproducible version of this figure.

TEACHER *Reflection*

What type of tracker can you and your colleagues create to prompt students to self-reflect on their learning after mid-unit common assessments?

_____

_____

_____

_____

_____

_____

_____

Self-Regulatory Feedback

Your collaborative team can also devise a system to help students generate their own self-regulatory feedback while taking a common formative check during a unit. Students not only solve the tasks on the assessment but also write the following next to each item.

- **Plus symbol:** Next to items if they feel confident about having solved well

- **Check mark:** Next to items if they feel partially confident about and wanted to see, using feedback, the aspects of the solution they should feel confident about and which parts might require additional learning

- **Question mark:** Next to items they were truly unsure about when taking the test but want feedback on to see if any part of their solution is on the right track

When students receive their assessments back, not only do they examine teacher feedback related to the mathematical tasks on the assessment but also check their solutions to grow their own self-regulatory feedback *during* the unit.

With the ease of scoring selected-response items using digital software, students should develop the same set of reflection skills when completing multiple-choice items.

As another example, any time you ask students to solve multiple-choice mathematics tasks, they might identify if they had solved them using one of the following four strategies.

1. Just do it (read the question, solve it, and find the answer)

2. Do it backwards (look at the solutions and substitute them in to find the answer)

3. Educated guess (eliminate one or more options and guess from there)

4. Pure guess (simply guess which option is the answer)

Next to each multiple-choice item, whether using clickers in class or on a common formative assessment, students wrote one, two, three, or four to generate self-regulatory feedback about their choices for solving problems. Ideally, students began to notice that strategies one and two should be chosen first to demonstrate learning and provide better feedback for continued learning. They also learned that strategy four meant even if every answer happened to be correct, they did not have evidence of having learned the essential learning standard. Additionally, this provided another opportunity for students to learn how to discuss solution pathways and plans with peers to learn from one another when giving each other feedback.

A key formative assessment feature used during the unit or after the unit ends will be the process you have in place for your students to use and respond to the feedback and results of their assessment performance. Use figure 6.4 to determine how your team will address feedback and student action during and after a unit from common formative assessments.

How will you help students identify strengths, weaknesses, and the essential learning standards students still need to meet? What structures will you use to measure their progress? What do you expect of students during the current unit or in the next unit to re-engage in learning, as needed? How will your team enrich or extend learning, as needed? These questions, when answered collectively as a team, and accompanied with quality feedback, guide the team to answer PLC critical questions 3 and 4 during the unit (DuFour et al., 2016): How will we respond when some students do not learn? and How will we extend the learning for students who are already proficient?

Directions: Use the following prompts to guide discussion of your current team agreements regarding feedback to students and student action to re-engage in learning.

1. How will we commit to calibrating our scoring and determine ways to give quality feedback to students?

2. What team structure or template will we commit to using with students to have them self-assess their learning and devise a learning plan, if needed, after the end-of-unit common assessment?

3. What team structure or template will we commit to using with students during a unit so they re-engage in learning by essential learning standard, as needed?

4. How will we require students to take action on their feedback in this unit or the next?

5. What strategies will we use to teach students how to give each other meaningful feedback and learn from one another?

Figure 6.4: Team discussion tool—Collaborative team feedback with student action.

*Visit **go.SolutionTree.com/MathematicsatWork** for a free reproducible version of this figure.*

TEACHER *Reflection*

How can you utilize mid-unit or other during-the-unit common mathematics assessments as part of your Tier 2 mathematics intervention program?

How can you utilize your mid-unit or during-the-unit common mathematics assessments to extend student learning during a unit?

_____

_____

_____

_____

_____

_____

_____

A strong formative feedback process requires students to use their common assessments for continued learning of essential learning standards. When done during the unit, gaps in student learning are more quickly closed, allowing students to be more prepared for learning in each subsequent unit.

TEAM RECOMMENDATION

Create a System for Student Action and Ownership of Learning Based on Formative Feedback *During* a Unit Using Mid-Unit Common Assessments

- Create a reflection tool for students to use throughout a unit as a way to monitor progress toward meeting essential learning standards before the unit ends.

- Require students to re-engage in learning if there are essential learning standards not yet understood.

- Determine how you can work together to minimize the number of students needing a Tier 2 intervention after the end-of-unit assessment by proactively working with students to learn the essential learning standards *during* the unit.

How can you and your colleagues ensure that every student closes learning gaps in mathematics? Are you willing to extend the mathematics learning for each student in your grade level or course through a strong formative assessment process? In such a process, you provide each student with more equitable learning expectations and the necessary scaffolding to meet those expectations.

Part 2 focused on the student actions required for continued learning based on the FAST feedback from your team-created common assessments (see page 67). It examined the actions and details of team action 2: Use common assessments for formative student learning and intervention.

Team action 2 helps you and your colleagues address PLC critical questions 3 and 4 (DuFour et al., 2016).

3. How will we respond when some students do not learn?

4. How will we extend the learning for students who are already proficient?

Recall the questions that started part 2: What happens in the classrooms across your collaborative team when you return a common mathematics assessment to your students? What do you expect students to do with the feedback you provide? An assessment instrument is meaningful and useful *only* when it becomes part of a formative assessment process of learning you and your colleagues establish.

The following four teacher actions are important in developing a quality formative assessment process.

1. Calibrate your scoring of common assessments by essential learning standard so students get FAST feedback that is consistent and from which they can learn.

2. Determine how students will self-reflect and set goals for learning based on feedback on common end-of-unit assessments.

3. Create a quality Tier 2 mathematics intervention program that requires students to re-engage in learning or extend just-in-time learning based on the essential learning standards.

4. Determine how students self-reflect and re-engage in learning using feedback from during-the-unit common assessments.

Taken together, your intentional common assessment design, calibrated scoring with feedback, and structures for student reflection and re-engagement in a strong intervention program create the formative assessment process linked to gains in student learning. Students are able to *reflect*, *refine*, and *act* over and over and then over again. In short, they practice learning mathematics.

Epilogue

By Timothy D. Kanold

An epilogue should serve as a conclusion to what has happened to you based on your experiences in reading and using this book. An epilogue should also serve as a conclusion to your growth and change about mathematics assessment as your teaching career unfolds.

During the 1990s and early 2000s, my colleagues and I at Adlai E. Stevenson High School District 125—the birthplace of the PLC at Work model—developed the deep mathematics assessment experiences revealed in this book. Stevenson eventually became a U.S. model for other schools and districts because of the deep inspection, revision, and eventual collective team response to the quality of the common unit-by-unit assessments in mathematics for each grade level and course.

At the time, many considered the decision to allow students to embrace their mistakes and use the results of unit mathematics assessments to monitor and improve their learning to be a groundbreaking practice. Yet once we collectively pursued this practice with certain ferocity, student performance began to soar. The talented coauthors of this book will tell you they experienced the same results in their schools and districts.

Developing and using high-quality mathematics assessments and requiring just-in-time student interventions necessary for learning each standard are essential responsibilities of our professional practice.

Just like an athletic coach measures success based on game-day results, our mathematics teacher teams at Stevenson began to base their success on game-day results from quizzes, tests, projects, and labs. For example, the five mathematics teachers on the algebra 1 team comprised the first teacher team at Stevenson (in any subject area or grade level) to take a more broad-based ownership of students enrolled in the course. Teachers were able to see beyond the students assigned only to them, and instead, saw themselves as coaches for *every student* in the course.

Teacher team members placed signs above their desks that read *algebra coach*. This allowed any algebra student in need of support to ask any of the algebra teachers, or coaches, for help (not just his or her own teacher).

This powerful teacher team worked together to develop common assessments, align the assessments around essential learning standards, collaboratively score those assessments, and reach agreements about the quality of student responses to assessment questions. It was the first team at Stevenson to benefit from a significant improvement in student learning.

The team established a common response to help all students embrace their mistakes and expected each student to re-engage in the essential learning standards for that unit, as necessary. None of these changes in practice were easy. In this book, we attempt to guide your conversations and your assessment work with colleagues to make the journey a bit easier.

On behalf of Sarah, Matt, Bill, Jessica, and Mona, may your assessment and intervention journey lead to inspired student learning each and every day.

Cognitive-Demand-Level Task Analysis Guide

This appendix provides criteria you can use to evaluate the cognitive-demand level of the tasks you choose each day. As you develop your unit plans, use this tool to ensure a balance of lower-level and higher-level tasks for students as they learn the standards.

Table A.1: Cognitive-Demand Levels of Mathematical Tasks

Lower-Level Cognitive Demand	Higher-Level Cognitive Demand
Memorization Tasks • These tasks involve reproducing previously learned facts, rules, formulae, or definitions to memory. • They cannot be solved using procedures because a procedure does not exist or because the time frame in which the task is being completed is too short to use the procedure. • They are not ambiguous; such tasks involve exact reproduction of previously seen material and what is to be reproduced is clearly and directly stated. • They have no connection to the concepts or meaning that underlie the facts, rules, formulae, or definitions being learned or reproduced.	**Procedures With Connections Tasks** • These procedures focus students' attention on the use of procedures for the purpose of developing deeper levels of understanding of mathematical concepts and ideas. • They suggest pathways to follow (explicitly or implicitly) that are broad general procedures that have close connections to underlying conceptual ideas as opposed to narrow algorithms that are opaque with respect to underlying concepts. • They usually are represented in multiple ways (for example, visual diagrams, manipulatives, symbols, or problem situations). They require some degree of cognitive effort. Although general procedures may be followed, they cannot be followed mindlessly. Students need to engage with the conceptual ideas that underlie the procedures in order to successfully complete the task and develop understanding.
Procedures Without Connections Tasks • These procedures are algorithmic. Use of the procedure is either specifically called for, or its use is evident based on prior instruction, experience, or placement of the task. • They require limited cognitive demand for successful completion. There is little ambiguity about what needs to be done and how to do it. • They have no connection to the concepts or meaning that underlie the procedure being used. • They are focused on producing correct answers rather than developing mathematical understanding. • They require no explanations or have explanations that focus solely on describing the procedure used.	**Doing Mathematics Tasks** • Doing mathematics tasks requires complex and no algorithmic thinking (for example, the task, instructions, or examples do not explicitly suggest a predictable, well-rehearsed approach or pathway). • It requires students to explore and understand the nature of mathematical concepts, processes, or relationships. • It demands self-monitoring or self-regulation of one's own cognitive processes. • It requires students to access relevant knowledge and experiences and make appropriate use of them in working through the task. • It requires students to analyze the task and actively examine task constraints that may limit possible solution strategies and solutions. • It requires considerable cognitive effort and may involve some level of anxiety for the student due to the unpredictable nature of the required solution process.

Source: © 1998 by Smith & Stein. Used with permission.

References and Resources

Aguirre, J., Mayfield-Ingram, K., & Martin, D. B. (2013). *The impact of identity in K–8 mathematics: Rethinking equity-based practices*. Reston, VA: National Council of Teachers of Mathematics.

Ainsworth, L. (2003). *"Unwrapping" the standards: A simple process to make standards manageable*. Englewood, CO: Advanced Learning Press.

AllThingsPLC. (n.d.). *Evidence of effectiveness: Field Elementary*. Accessed at www.allthingsplc.info/evidence/details/id,998 on September 19, 2017.

Bailey, K., & Jakicic, C. (2012). *Common formative assessment: A toolkit for Professional Learning Communities at Work*. Bloomington, IN: Solution Tree Press.

Black, P., & Wiliam, D. (2001). *Inside the black box: Raising standards through classroom assessment*. London: Assessment Group of the British Educational Research Association.

Buffum, A., Mattos, M., & Malone, J. (2018). *Taking action: A handbook for RTI at Work™*. Bloomington, IN: Solution Tree Press.

Chappuis, S. C., & Stiggins, R. J. (2002). Classroom assessment for learning. *Educational Leadership, 60*(1), 40–43.

Covey, S. R. (1989). *The seven habits of highly effective people: Restoring the character ethic*. New York: Simon and Schuster.

Covey, S. R. (2004). The 7 habits of highly effective people: Restoring the character ethic (Rev. ed.). New York: Free Press.

DuFour, R. (2015). *In praise of American educators: And how they can become even better*. Bloomington, IN: Solution Tree Press.

DuFour, R., DuFour, R., Eaker, R., & Karhanek, G. (2009). *Raising the bar and closing the gap: Whatever it takes*. Bloomington, IN: Solution Tree Press.

DuFour, R., DuFour, R., Eaker, R., Many, T. W., & Mattos, M. (2016). *Learning by doing: A handbook for Professional Learning Communities at Work* (3rd ed.). Bloomington, IN: Solution Tree Press.

DuFour, R., & Marzano, R. J. (2011). *Leaders of learning: How district, school, and classroom leaders improve student achievement*. Bloomington, IN: Solution Tree Press.

Eaker, R., & Keating, J. (2012). *Every school, every team, every classroom: District leadership for growing Professional Learning Communities at Work*. Bloomington, IN: Solution Tree Press.

Erkens, C., Schimmer, T., & Vagle, N. D. (2017). *Essential assessment: Six tenets for bringing hope, efficacy, and achievement to the classroom*. Bloomington, IN: Solution Tree Press.

Fernandes, A., Crespo, S., & Civil, M. (2017). Introduction. In A. Fernandes, S. Crespo, & M. Civil (Eds.), *Access and equity: Promoting high-quality mathematics, grades 6–8* (pp. 1–10). Reston, VA: National Council of Teachers of Mathematics.

Guskey, T. R. (Ed.). (2009). *The teacher as assessment leader*. Bloomington, IN: Solution Tree Press.

Hansen, A. (2015). *How to develop PLCs for singletons and small schools*. Bloomington, IN: Solution Tree Press.

Hattie, J. (2009). *Visible learning: A synthesis of over 800 meta-analyses relating to achievement*. New York: Routledge.

Hattie, J. (2012). *Visible learning for teachers: Maximizing impact on learning*. New York: Routledge.

Hattie, J., Fisher, D., & Frey, N. (2017). *Visible learning for mathematics: What works best to optimize student learning*. Thousand Oaks, CA: Corwin Press.

Jilk, L. M., & Erickson, S. (2017). Shifting students' beliefs about competence by integrating mathematics strengths into tasks and participation norms. In A. Fernandes, S. Crespo, & M. Civil (Eds.), *Access and equity: Promoting high-quality mathematics, grades 6–8* (pp. 11–26). Reston, VA: National Council of Teachers of Mathematics.

Kanold, T. D., Barnes, B., Larson, M. R., Kanold-McIntyre, J., Schuhl, S., & Toncheff, M. (in press). *Mathematics homework and grading in a PLC at Work*. Bloomington, IN: Solution Tree Press.

Kanold, T. D. (Ed.), Briars, D. J., Asturias, H., Foster, D., & Gale, M. A. (2013). *Common Core mathematics in a PLC at Work, grades 6–8*. Bloomington, IN: Solution Tree Press.

Kanold, T. D. (Ed.), Dixon, J. K., Adams, T. L., & Nolan, E. C. (2015). *Beyond the Common Core: A handbook for mathematics in a PLC at Work, grades K–5*. Bloomington, IN: Solution Tree Press.

Kanold, T. D., Kanold-McIntyre, J., Larson, M. R., Barnes, B., Schuhl, S., & Toncheff, M. (in press). *Mathematics instruction and tasks in a PLC at Work*. Bloomington, IN: Solution Tree Press.

Kanold, T. D. (Ed.), Kanold-McIntyre, J., Larson, M. R., & Briars, D. J. (2015). *Beyond the Common Core: A handbook for mathematics in a PLC at Work, grades 6–8*. Bloomington, IN: Solution Tree Press.

Kanold, T. D. (Ed.), & Larson, M. R. (2012). *Common Core mathematics in a PLC at Work, leader's guide*. Bloomington, IN: Solution Tree Press.

Kanold, T. D. (Ed.), & Larson, M. R. (2015). *Beyond the Common Core: A handbook for mathematics in a PLC at Work, leader's guide*. Bloomington, IN: Solution Tree Press.

Kanold, T. D. (Ed.), Larson, M. R., Fennell, F., Adams, T. L., Dixon, J. K., Kobett, B. M., & Wray, J. A. (2012a). *Common Core mathematics in a PLC at Work, grades K–2*. Bloomington, IN: Solution Tree Press.

Kanold, T. D. (Ed.), Larson, M. R., Fennell, F., Adams, T. L., Dixon, J. K., Kobett, B. M., & Wray, J. A. (2012b). *Common Core mathematics in a PLC at Work, grades 3–5*. Bloomington, IN: Solution Tree Press.

Kanold, T. D. (Ed.), & Toncheff, M. (2015). *Beyond the Common Core: A handbook for mathematics in a PLC at Work, high school*. Bloomington, IN: Solution Tree Press.

Kanold, T. D., Toncheff, M., Larson, M. R., Barnes, B., Kanold-McIntyre, J., & Schuhl, S. (in press). *Mathematics coaching and collaboration in a PLC at Work*. Bloomington, IN: Solution Tree Press.

Kanold, T. D. (Ed.), Zimmermann, G., Carter, J. A., Kanold, T. D., & Toncheff, M. (2012). *Common Core mathematics in a PLC at Work, high school*. Bloomington, IN: Solution Tree Press.

Kramer, S. V. (2015). *How to leverage PLCs for school improvement*. Bloomington, IN: Solution Tree Press.

Mattos, M., DuFour, R., DuFour, R., Eaker, R., & Many, T. W. (2016). *Concise answers to frequently asked questions about Professional Learning Communities at Work*. Bloomington, IN: Solution Tree Press.

Missoula County Public Schools Curriculum Consortium. (2013, June). *PreK–12 mathematics curriculum*. Accessed at www.mcpsmt.org/cms/lib03/MT01001940/Centricity/Domain/825/Mathematics%20Curriculum.pdf on September 27, 2017.

National Council of Supervisors of Mathematics. (n.d.). *Common Core State Standards: Illustrating the Standards for Mathematical Practice—Module index*. Accessed at www.mathedleadership.org/ccss/itp/index.html on September 19, 2017.

National Council of Supervisors of Mathematics. (2015). *Improving student achievement by infusing highly effective instructional strategies into Multi-Tiered Support Systems (MTSS)–Response to Intervention (RtI) Tier 2 instruction*. Accessed at www.mathedleadership.org/member/docs/resources/positionpapers/NCSMPositionPaper15.pdf on June 1, 2017.

National Council of Teachers of Mathematics. (2014). *Principles to actions: Ensuring mathematical success for all*. Reston, VA: Author.

National Governors Association Center for Best Practices & Council of Chief State School Officers. (2010). *Common Core State Standards for mathematics*. Washington, DC: Authors. Accessed at www.corestandards.org/assets/CCSSI_Math%20Standards.pdf on October 5, 2017.

The National Science Digital Library. (n.d.). *Search results: mathematics*. Accessed at https://nsdl.oercommons.org/browse?f.general_subject=mathematics on September 19, 2017.

Popham, W. J. (2011). *Transformative assessment in action: An inside look at applying the process*. Alexandria, VA: Association for Supervision and Curriculum Development.

Reeves, D. (2011). *Elements of grading: A guide to effective practice*. Bloomington, IN: Solution Tree Press.

Reeves, D. (2016). *Elements of grading: A guide to effective practice* (2nd ed.). Bloomington, IN: Solution Tree Press.

Reeves, D. B. (2002). *The leader's guide to standards: A blueprint for educational equity and excellence*. San Francisco: Jossey-Bass.

Schimmer, T. (2016). *Grading from the inside out: Bringing accuracy to student assessment through a standards-based mindset*. Bloomington, IN: Solution Tree Press.

Smarter Balanced Assessment Consortium. (n.d.). *Sample index*. Accessed at http://sampleitems.smarterbalanced.org /BrowseItems on September 19, 2017.

Smith, M. S., & Stein, M. K. (1998). Selecting and creating mathematical tasks: From research to practice. *Mathematics Teaching in the Middle School, 3*(5), 348.

Smith, M. S., & Stein, M. K. (2011). *5 practices for orchestrating productive mathematics discussions*. Reston, VA: National Council of Teachers of Mathematics.

Smith, M. S., Steele, M. D., & Raith, M. L. (2017). *Taking action: Implementing effective mathematics teaching practices, grades 6–8*. Reston, VA: National Council of Teachers of Mathematics.

Stiggins, R. (2007). Assessment through the student's eyes. *Educational Leadership, 64*(8), 22–26. Accessed at www.ascd.org /publications/educational-leadership/may07/vol64/num08/Assessment-Through-the-Student's-Eyes.aspx on September 19, 2017.

Stiggins, R., Arter, J., Chappuis, J., & Chappuis, S. (2006). *Classroom assessment* for *student learning: Doing it right—using it well*. Princeton, NJ: Educational Testing Service.

Wiliam, D. (2011). *Embedded formative assessment*. Bloomington, IN: Solution Tree Press.

Wiliam, D. (2016). The secret of effective feedback. *Educational Leadership, 73*(7), 10–15.

Wiliam, D. (2018). *Embedded formative assessment* (2nd ed.). Bloomington, IN: Solution Tree Press.

Index

Every Student Can Learn Mathematics series
Timothy D. Kanold et al.

Discover a comprehensive PLC at Work® approach to achieving mathematics success in K–12 classrooms. Each book offers two teacher team or coaching actions that empower teams to reflect on and refine current practices and routines based on high-quality, research-affirmed criteria.

BKF823 BKF824 BKF825 BKF826

Mathematics Instruction and Tasks in a PLC at Work®
Timothy D. Kanold, Jessica Kanold-McIntyre, Matthew R. Larson, Bill Barnes, Sarah Schuhl, Mona Toncheff

Improve your students' comprehension and perseverance in mathematics. This user-friendly resource will help you identify high-quality lesson-design elements and then show you how to implement them within your classroom. The book features sample lesson templates, online resources for instructional support, and more.

BKF824

Mathematics Homework and Grading in a PLC at Work®
Timothy D. Kanold, Bill Barnes, Matthew R. Larson, Jessica Kanold-McIntyre, Sarah Schuhl, Mona Toncheff

Rely on this user-friendly resource to help you create common independent practice assignments and equitable grading practices that boost student achievement in mathematics. The book features teacher team tools and activities to inspire student achievement and perseverance.

BKF825

Mathematics Coaching and Collaboration in a PLC at Work®
Timothy D. Kanold, Mona Toncheff, Matthew R. Larson, Bill Barnes, Jessica Kanold-McIntyre, Sarah Schuhl

Build a mathematics teaching community that promotes learning for K–12 educators and students. This user-friendly resource will help you coach highly effective teams within your PLC and then show you how to utilize collaboration and lesson-design elements for team reflection, data analysis, and action.

BKF826

GL⬤BAL PD

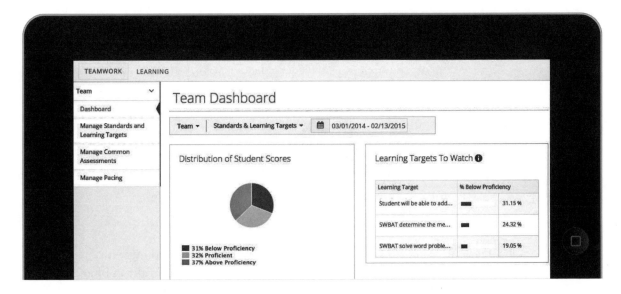

The **Power to Improve**
Is in Your Hands

Global PD gives educators focused and goals-oriented training from top experts. You can rely on this innovative online tool to improve instruction in every classroom.

- Get unlimited, on-demand access to guided video and book content from top Solution Tree authors.

- Improve practices with personalized virtual coaching from PLC-certified trainers.

- Customize learning based on skill level and time commitments.

Solution Tree's mission is to advance the work of our authors. By working with the best researchers and educators worldwide, we strive to be the premier provider of innovative publishing, in-demand events, and inspired professional development designed to transform education to ensure that all students learn.

The National Council of Teachers of Mathematics is a public voice of mathematics education, supporting teachers to ensure equitable mathematics learning of the highest quality for all students through vision, leadership, professional development, and research.